Monographs in Theoretical Computer Science
An EATCS Series

Springer
Berlin
Heidelberg
New York
Hong Kong
London
Milan
Paris
Tokyo

Lane A. Hemaspaandra · Leen Torenvliet

Theory of
Semi-Feasible Algorithms

With 6 Figures

Springer

Authors

Prof. Dr. Lane A. Hemaspaandra
Department of Computer Science
University of Rochester
Rochester, NY 14627
USA
lane@cs.rochester.edu

Dr. Leen Torenvliet
Department of Computer Science
University of Amsterdam
Plantage Muidergracht 24
1018 TV Amsterdam
The Netherlands
leen@science.uva.nl

Series Editors

Prof. Dr. Wilfried Brauer
Institut für Informatik
Technische Universität München
Arcisstrasse 21, 80333 München, Germany
brauer@informatik.tu-muenchen.de

Prof. Dr. Grzegorz Rozenberg
Leiden Institute of Advanced Computer Science
University of Leiden
Niels Bohrweg 1, 2333 CA Leiden, The Netherlands
rozenber@liacs.nl

Prof. Dr. Arto Salomaa
Data City
Turku Centre for Computer Science
20 500 Turku, Finland
asalomaa@utu.fi

Library of Congress Cataloging-in-Publication Data

Hemaspaandra, Lane A.
    Theory of semi-feasible algorithms/Lane A. Hemaspaandra, Leen Torenvliet.
        p. cm.
    Includes bibliographical references and index.

    1. Computer algorithms. 2. Computational complexity. I. Torenvliet, Leen. II. Title.

QA76.9.A43 H46 2002
005.1–dc21

2002029158

ACM Computing Classification (1998): F.1.1, F.1.2, F.1.3, F.2.2, F.2.3, I.2.8

ISBN 978-3-642-07581-0

Springer-Verlag Berlin Heidelberg New York,
a member of BertelsmannSpringer Science+Business Media GmbH

© Springer-Verlag Berlin Heidelberg 2010
Printed in Germany

Cover Design: KünkelLopka, Heidelberg

Preface

An Invitation to the Dance

It is an underappreciated fact that sets may have various types of complexity, and not all types are in harmony with each other. The primary goal of this book is to unify and make more widely accessible a vibrant stream of research—the theory of semi-feasible computation—that perfectly showcases the richness of, and contrasts between, the central types of complexity.

The semi-feasible sets, which are most commonly referred to as the P-selective sets, are those sets L for which there is a deterministic polynomial-time algorithm that, when given as input any two strings of which at least one belongs to L, will output one of them that is in L. The reason we say that the semi-feasible sets showcase the contrasts among types of complexity is that it is well-known that many semi-feasible sets have no recursive algorithms (thus their time complexity cannot be upper-bounded by standard time-complexity classes), yet all semi-feasible sets are simple in a wide range of other natural senses. In particular, the semi-feasible sets have small circuits, they are in the extended low hierarchy, and they cannot be NP-complete unless $P = NP$.

The semi-feasible sets are fascinating for many reasons. First, as mentioned above, they showcase the fact that mere deterministic time complexity is not the only potential type of complexity in the world of computation. Sets that are complex in terms of deterministic time—such as nonrecursive P-selective sets—may nonetheless be simple in many other computationally natural senses. A second reason that the semi-feasible sets are interesting is that they crisply capture the complexity of (standard left cuts of) real numbers, and a recent refinement of the semi-feasible sets has been shown to capture the complexity of complexity-bounded real numbers.

A third and more historical reason for interest in the semi-feasible sets is that they form the complexity-theoretic analog of a key class from recursive function theory; the semi-feasible sets are exactly what one gets if one alters the definition of the semi-recursive sets by changing the selector function from "recursive" to "polynomial-time computable." In the late 1960s, the semi-recursive sets yielded great insights into distinguishing the power of reductions in the recursion-theoretic context. In 1979, Alan Selman launched a program that used—successfully, in the context of structural connections to exponential time—semi-feasible sets to understand the structure of polynomial-time reductions.

A fourth and somewhat surprising reason to study semi-feasible sets is that the semi-feasible sets (in their nondeterministic version) conditionally resolve the important issue of whether NP machines can cull down to one the large number of potential solutions of satisfiable formulas. In particular, the study of the semi-feasible sets has established (see Section 2.4) that NP lacks such "unique solutions" unless the polynomial hierarchy collapses.

A fifth reason to study the semi-feasible sets is that the notion of semi-feasibility is both natural and attractive, and fits well into two related broad themes of computer science: making computers "smarter" even on problems that may be too complex to solve exactly, and allowing computers to make decisions even when they lack absolute "knowledge" of the goodness of the choices involved. For computers to be able to act and interact more intelligently with users, thus helping make computing more intuitive to those users, it would be nice for the computers themselves to show some "intuition" when making decisions, i.e., to act more boldly and intuitively—perhaps making membership claims that they might not "know" to be absolutely right or wrong, but that merely skate on intuition. Selectivity theory studies the sets for which a polynomial-time algorithm given two inputs—viewed as options, and potentially even as actions toward some goal—can intuit one to try, i.e., one for which to say "yes, if I had to take a flier and declare one of those options to be a good one, I'd go with *this* one." Algorithms satisfying the rules of selectivity will have the property that if there is any good choice—one having whatever properties are possessed by the options in the set—offered to them, they will make a good choice. Curiously enough, due to the possibility of there being no good choice among the options being considered, or there being no bad choice among the options being considered, the algorithms will not necessarily "know" whether their choice is good or whether any option they pass up is bad. Nonetheless, we know that they are acting intelligently: If there was a good option among the inputs, a good option was chosen. Thus, by studying selectivity theory, we study the extent to which polynomial-time decision-making can be made "smart." Speaking more broadly, one may say that selectivity theory formalizes a natural notion of intuition and intuitive computing.

We feel that this is the right time for such a book as this. Research into semi-feasible computation has already developed a rich set of tools, yet is young enough to have an abundance of fresh open issues. Though the primary goal of this book is to unify semi-feasibility research and make it accessible, another major goal is to lay out a path along which the reader can meet and engage the open problems in this research area. And wonderful open problems do remain. Though during the past fifteen years many long-standing issues were resolved, and the semi-feasible sets were shown to be deeply connected to issues of uniqueness, self-reducibility, and nondeterminism, these very advances themselves motivated new questions. The confluence of exciting open issues and a rich and expanding set of technical tools with which to study the semi-feasible sets make this perhaps the best of times to join the search for knowledge about semi-feasible computation. We hope this book will serve as both an invitation and a pathway.

Logistics

No previous knowledge of semi-feasible computation is required to read this book. We start with the definition of semi-feasibility and move on from there. However, though we include in the text or the appendix full definitions of each complexity-theoretic notion the book uses, we do assume that the reader has the basic comfort with computational complexity concepts—and the ability to grasp new definitions—that one would gain from a typical first course on computational complexity theory. (Among the textbooks, at various levels of difficulty, on computational complexity are those of Balcázar, Díaz, and Gabarró [BDG95,BDG90], Bovet and Crescenzi [BC93], Du and Ko [DK00], Hemaspaandra and Ogihara [HO02], Homer and Selman [HS01], Papadimitriou [Pap94], and Sipser [Sip97, Part Three]).

This text can be the focus of a second course on computational complexity theory. In particular, we feel that this material is very appropriate as a seminar course for first- or second-year graduate students who have already taken a first computational complexity course. We have found that both theory and non-theory students value and much enjoy the concreteness and "tour of the cutting edge" aspects of a course devoted to semi-feasible computation.

In virtually all of Chapters 1 through 6, the text contains no citations. The citations in these sections can be found in the Bibliographic Notes sections that end each chapter. The "we" used in this book (e.g., "we define," "we prove") refers to the reader and the authors as we together explore the theory of semi-feasible computation. Nonetheless, some of the research this book covers was done by the authors and their coauthors, and we sincerely thank those coauthors with whom we have explored semi-feasible computation: E. Allender, H. Buhrman, P. van Emde Boas, E. Hemaspaandra, H. Hempel, A. Hoene, Z. Jiang, A. Naik, C. Nasipak, A. Nickelsen, M. Ogihara, K. Parkins, J. Rothe, A. Selman, T. Thierauf, J. Wang, O. Watanabe, M. Zaki, and M. Zimand. Such research was generously funded by the following grants, whose support we gratefully acknowledge: HC&M-ERB4050PL93-0516, NSF-CCR-8957604, NSF-INT-9116781/ JSPS-ENGR-207, NSF-CCR-9322513, NSF-INT-9513368/DAAD-315-PRO-fo-ab, NSF-INT-9815095/DAAD-315-PPP-gü-ab, and NWO-R-62-561.

We are extremely grateful to C. Homan, T. Tantau, and M. Thakur for proofreading the entire book, and to W. Gasarch, M. de Graaf, S. Homer, K. Regan, J. Rothe, D. Sivakumar, M. Stol, and J. Verbeek, each of whom did a detailed proofreading of one or more chapters of an earlier draft of this book. This book benefited greatly from their suggestions and insights. We also thank the many other people who helped us with advice, discussions, suggestions, most-recent-version-of-paper information, or literature pointers, including E. Allender, H. Buhrman, J. Cai, L. Fortnow, E. Hemaspaandra, G. Magklis, A. Nickelsen, M. Ogihara, and F. Veltman. We are grateful to the Springer series editors—W. Brauer, G. Rozenberg, and A. Salomaa—and staff—A. Hofmann, F. Holzwarth, U. Stricker, T. Toomey, H. Wössner, and especially I. Mayer—for their advice and help.

Above all, we thank our families for their love and encouragement.

Rochester, New York, September 2002   *Lane A. Hemaspaandra*
Amsterdam, September 2002   *Leen Torenvliet*

Contents

1. Introduction to Semi-Feasible Computation

1.1 P-Selectivity

1.1.1 Background and Definitions

Much of complexity theory focuses on the *membership complexity* of sets. The central question in membership complexity is, for a fixed set A: How hard is it to test whether an element is a member of the set? In particular, is there a polynomial-time algorithm that tests for membership in A? If so, we say A is a *feasible* set or, equivalently, A is in the complexity class P.

In 1979, Alan Selman proposed the complexity-theoretic study of a different property of sets: their *semi-membership complexity*. The central question in semi-membership complexity is, for a fixed set A: How hard is it to determine which of two given elements is "more likely" (or, more precisely, "logically no less likely") than the other to belong to A? We mean this in the sense of always choosing one of any two given inputs, and if exactly one of the inputs is in A then that is the one chosen. In particular, is there a polynomial-time function doing such choosing for A? If so, we say that A is a *semi-feasible* set or, equivalently, is a *P-selective* set.

More formally, the definition of semi-feasibility is as follows. Here, as elsewhere in the book unless otherwise explicitly stated or implicit from context, our alphabet is $\Sigma = \{0, 1\}$ and our sets are subsets of Σ^*.

Definition 1.1

1. *A set A is* semi-feasible *(equivalently,* P-selective*) if there is a (total) polynomial-time computable function f such that, for each $x, y \in \Sigma^*$,*
   a) $f(x, y) = x$ or $f(x, y) = y$, and
   b) if $x \in A$ or $y \in A$, then $f(x, y) \in A$.
   We say that such an f is a P-selector function *for A.*
2. P-sel $= \{A \,|\, A$ *is semi-feasible*$\}$.

Thus, a set is P-selective if there is a two-argument polynomial-time computable function that always outputs one of its arguments that is "logically no less likely" than the other argument to belong to the set.

There are a number of motivations for the study of P-selectivity. Historically, the semi-feasible sets are the complexity-theoretic analog of the semi-recursive sets from recursive function theory. Selman's immediate goal in introducing the P-selective sets was to distinguish the relative powers of different polynomial-time reductions in the context of NP sets. For example, he was interested in whether the assumption P $\neq$ NP is sufficient to ensure that $\leq_m^p$ reductions and $\leq_T^p$ reductions differ on NP sets. The reason Selman hoped that the P-selective sets would prove useful in this context is that their recursion-theoretic analogs had been central in separating the power of reducibilities when studying reductions relating recursively enumerable sets.

Later work (see Theorem 4.9) made it clear that the P-selective sets cannot distinguish between NP-$\leq_m^p$-completeness and NP-$\leq_T^p$-completeness unless the polynomial hierarchy collapses. Nonetheless, Selman successfully used the P-selective sets to distinguish between different kinds of polynomial-time reducibilities, though under complexity-theoretic assumptions that are substantially stronger than one might have hoped. Corollaries 1.17 and 1.18 state typical results from this research line.

Semi-feasible computation has a resonance far richer than merely that of distinguishing reducibilities. Even in the late 1970s, computer scientists were already deeply troubled by the fact that, although polynomial-time computation was widely accepted as the most natural notion of feasible computation, many crucial sets were not known to have polynomial-time membership algorithms. One reaction to this was to define and study new complexity classes to capture a variety of notions of "almost" polynomial time. Among such notions that have, then or since, been defined are the following.

1. P-sel, the P-selective sets—those sets having polynomial-time semi-membership algorithms.
2. P-close, the P-close sets—those sets having sparse[1] symmetric difference with some set in P.
3. NT, the near-testable sets—those sets having a polynomial-time algorithm determining, on input x, whether exactly one of x and x's lexicographical predecessor is in the set.
4. NNT, the implicitly membership-testable sets—those sets A having a polynomial-time algorithm that on each input $x \neq \epsilon$ correctly prints one of the following statements:
   a) "$x \in A$."
   b) "$x \notin A$."
   c) "$\|\{x, predecessor(x)\} \cap A\| = 1$."
   d) "$\|\{x, predecessor(x)\} \cap A\| \equiv 0 \pmod 2$."
5. qP, the quasipolynomial-time sets—those sets in $\bigcup_{k>0} \text{DTIME}[2^{\log^k n}]$.
6. APT, the almost polynomial-time sets—those sets accepted by some deterministic machine that runs in polynomial time on all but a sparse set of inputs.

---

[1] A set A is *sparse* if for some polynomial q it holds that, for each n, $\|A^{=n}\| \leq q(n)$.

Of these notions, P-selectivity has been by far the most intensely studied. Though this is in part due to the fact that the concept of semi-membership complexity is itself quite natural, it is also due to the fact that the P-selective sets satisfy many of complexity theory's classic "simplicity" tests, such as having small circuits/small advice (Chapter 2), being in the low hierarchy (Chapter 3), and not being hard for complexity classes unless unexpected complexity class collapses occur (Chapter 4).

One natural way to judge whether a concept is worthy of study is whether it yields interesting results even on topics beyond those that the concept would seem to directly address. Selectivity, especially in the nondeterministic analog we will discuss in the following section (Section 1.2), has exactly this property; it solves interesting, fundamental problems that are seemingly unrelated to selectivity theory. Section 2.4 provides a good example of this, namely, that selectivity is critical in establishing the current understanding on whether NP has "unique solutions." That isn't the only example of a broader use of selectivity theory. Selman's original motivation, which we mentioned earlier, is of this sort also, i.e., many of the applications of selectivity to understanding reductions have the property that selectivity is used "under the hood," rather than appearing explicitly in the result (see, e.g., Corollary 1.17).

Finally, we mention the classic example of P-selectivity: standard left cuts of real numbers. For any real number r, $0 \leq r < 1$, define

$$left(r) = \{b_1 b_2 b_3 \cdots b_z \mid (\forall j : 1 \leq j \leq z)\, [b_j \in \{0,1\}] \wedge r > \sum_{1 \leq i \leq z} \frac{b_i}{2^i}\}.$$

That is, the standard left cut of r is the set of nonnegative dyadic rationals less than r. For any real number r, $0 \leq r < 1$, the standard left cut of r is P-selective.

Theorem 1.2 *Every standard left cut is a* P*-selective set. That is, for each* r *satisfying* $0 \leq r < 1$, *it holds that* $left(r) \in$ P-sel.

Proof Consider the P-selector function f such that $f(a,b) = a$ if $a < b$ and otherwise $f(a,b) = b$, where "$a < b$" denotes that, with both viewed as dyadic rational fractions, a is less than b. Since a smaller number is no less likely to be less than r than a larger number, this P-selector will always choose a member of $left(r)$ if either input is in $left(r)$.   □

In fact, Theorem 1.2 is a reflection of a more general behavior that we will state as Theorem 1.15.

1.1.2 Basic Properties

In this section, we prove some basic properties of the P-selective sets. Some will be useful in future proofs, and others serve both to give a concrete intuition for proofs about P-selectivity and to present some of the "classic" results that spawned research lines that will be studied in depth in later chapters.

We state as a proposition the following fact, whose truth is immediately clear since, for a set $A \in$ P, one may use the polynomial-time selector function f_A defined by: $f_A(x, y)$ equals x if $x \in A$ and otherwise $f_A(x, y)$ equals y.

Proposition 1.3 P $\subseteq$ P-sel.

Though P-selector functions (Definition 1.1) for P-selective sets may be sensitive to the order of their two arguments, every P-selective set has *some* P-selector function that is oblivious to the order of its arguments, i.e., is a symmetric function. This easy observation will be used repeatedly in this book, since it simplifies proofs by allowing arguments to be viewed as unordered pairs rather than as ordered pairs.

Theorem 1.4 *If A is P-selective then A is P-selective via some P-selector function f satisfying*

$$(\forall x, y)\,[f(x, y) = f(y, x)].$$

Proof Let A be P-selective via P-selector function f'. Let $f(x, y) = f'(x, y)$ if $f'(x, y) = f'(y, x)$ and let $f(x, y) = \min(x, y)$ otherwise. Clearly, f is a P-selector for A and satisfies $(\forall x, y)\,[f(x, y) = f(y, x)]$.   ❑

Selman's seminal paper on P-selectivity proved (among many other things) the following claim about "self-reducibility," a topic that we will study in Sections 5.3 and 5.4.

Theorem 1.5

1. *For all sets A, $\emptyset \neq A \neq \Sigma^*$, it holds that $A \in$ P if and only if*

$$\overline{A} \leq_m^p A \text{ and } A \text{ is P-selective.}$$

2. *If A is P-selective and $B \leq_m^p A$ then B is P-selective.*

Part 2 of Theorem 1.5 has since been extended, as we will see in Section 5.3, to the case of positive Turing reductions. However, for completeness, and due to their directness, clarity, and simplicity, we include here a proof of these useful facts.

Proof of Theorem 1.5 We first prove part 1. Let A be a set other than $\emptyset$ and Σ^*. If $A \in$ P, then clearly A is P-selective (via $f(x, y) = x$ if $x \in A$ and $f(x, y) = y$ otherwise) and $\overline{A} \leq_m^p A$ (via the reduction $g(x) = in$ if $x \notin A$ and $g(x) = out$ if $x \in A$, where in and out are fixed elements respectively in A and $\overline{A}$). Conversely, if $\overline{A} \leq_m^p A$ via the polynomial-time function g and A is P-selective via P-selector function f, then $A \in$ P since in this case clearly $x \in A$ if and only if $f(x, g(x)) = x$, and this can be tested in polynomial time.

We now prove part 2. If A is P-selective via P-selector function f and $B \leq_m^p A$ via (polynomial-time) reduction g, then B is P-selective via the P-selector function $f_g(x, y)$ that equals x if $g(x) = f(g(x), g(y))$ and that equals y otherwise.   ❑   Theorem 1.5

Throughout this book, by NP-complete we mean NP-$\leq^p_m$-complete. Selman's seminal paper proved that no NP-complete set is P-selective unless P = NP. This basic result, whose proof is included below as Theorem 1.6, spawned the elaborate research line covered in Chapter 4. The proof below exploits the *2-disjunctive self-reducibility* of SAT, the set of satisfiable boolean formulas. A set A is *Turing self-reducible* if there is a deterministic polynomial-time Turing machine M such that $A = L(M^A)$ and, for each x, $M^A(x)$ queries only strings of lengths strictly less than $|x|$. If the acceptance behavior of M is such that on each input M accepts exactly when M either (i) asks at least one query that is in the oracle set or (ii) asks no queries of the oracle and halts in an accepting state, then we say that A is *disjunctively self-reducible*. If A is disjunctively self-reducible via a (polynomial-time) machine M that on each input asks at most two oracle questions, then we say that A is *2-disjunctively self-reducible*.

Theorem 1.6 *If there exists an NP-complete P-selective set, then* P = NP.

Proof Let A be NP-complete and P-selective. So SAT $\leq^p_m A$. Note that $\emptyset \neq A \neq \Sigma^*$, since neither $\emptyset$ nor Σ^* is NP-complete (even if P = NP). So, by Theorem 1.5, SAT is P-selective. Let f be a P-selector function for SAT. We give a deterministic polynomial-time algorithm for SAT. On input F, without loss of generality, let the variables be, for some k, $v_1, v_2, \ldots, v_k$ (otherwise rename the variables so that this holds). If $k = 0$ we are done; F has no variables and can be evaluated in polynomial time. Otherwise, proceed as follows. Let F_{list} denote F with the substitutions listed in *list* performed. For example, $F_{v_k=1}$ denotes F with v_k assigned the value true (in such lists, 1 will denote true and 0 will denote false). Run $f(F_{v_k=1}, F_{v_k=0})$. Note that

$$F \in \text{SAT} \iff f(F_{v_k=1}, F_{v_k=0}) \in \text{SAT}.$$

If $k = 1$ then we are done; $f(F_{v_k=1}, F_{v_k=0})$ has no unassigned variables and can be evaluated in polynomial time. If $k > 1$, let $b_k = 1$ if $f(F_{v_k=1}, F_{v_k=0}) = F_{v_k=1}$ and otherwise let $b_k = 0$. Run $f(F_{v_k=b_k, v_{k-1}=1}, F_{v_k=b_k, v_{k-1}=0})$. Note that

$$F \in \text{SAT} \iff f(F_{v_k=b_k, v_{k-1}=1}, F_{v_k=b_k, v_{k-1}=0}) \in \text{SAT}.$$

If $k = 2$ we are done via evaluating $f(F_{v_k=b_k, v_{k-1}=1}, F_{v_k=b_k, v_{k-1}=0})$, which has no unassigned variables. Otherwise, continue in a similar fashion. After at most k applications of f to pairs of formulas each no longer than the original input, we have correctly determined whether $F \in \text{SAT}$. $\qquad\square$

Corollary 1.7 *If* NP $\subseteq$ P-sel *then* P = NP.

Selman mentioned in his original papers on P-selectivity that the P-selective sets are closed under complementation.

Theorem 1.8 $A \in$ P-sel $\iff \overline{A} \in$ P-sel.

This is clear, since if f' is a P-selector function for A, then

$$f(x,y) = \begin{cases} y & \text{if } f'(x,y) = x, \\ x & \text{otherwise} \end{cases}$$

is a P-selector function for $\overline{A}$. In fact, as we will see in Chapter 5, the closure properties of the P-selective sets have since been much studied. In particular, we will see that of the 2^{2^k} k-ary boolean functions, exactly $2k + 2$ are closure properties of the P-selective sets, and the remaining $2^{2^k} - 2k - 2$ are properties under which the P-selective sets are not closed.

In the previous section, standard left cuts were used as examples of P-selective sets. In fact, this reflects the more general behavior that each initial segment of a polynomial-time computable linear ordering of Σ^*, $\prec$, is itself a P-selective set. Note that the notion of "initial segment" used here—basically, being closed downward under $\prec$—does allow even infinite sets to be initial segments.

Definition 1.9 *Let $\prec$ denote a linear ordering of Σ^*. A set A is said to be an* initial segment *of the linear ordering $(\Sigma^*, \prec)$ if for all x and y it holds that*

$$(x \in A \wedge y \prec x) \implies y \in A.$$

Theorem 1.10 *If A is an initial segment of a polynomial-time computable linear ordering of Σ^*, then A is P-selective.*

The proof of Theorem 1.10 is immediate. The P-selector function $f(x,y)$ outputs x if x is less than or equal to y with respect to the ordering, and otherwise outputs y. Theorem 1.10 is an implication rather than a complete characterization. This is somewhat curious since the recursion-theoretic analog is a complete characterization.

Definition 1.11 *A set A is* semi-recursive *if there is a recursive function f such that, for each $x, y \in \Sigma^*$,*

1. *$f(x,y) = x$ or $f(x,y) = y$, and*
2. *if $x \in A$ or $y \in A$, then $f(x,y) \in A$.*

Proposition 1.12 *A is semi-recursive if and only if A is an initial segment of some recursive linear ordering of Σ^*.*

In fact, complete complexity-theoretic characterizations can be obtained either by focusing only on tally sets (Theorem 1.13) or by modifying the approach to ordering (Theorem 1.15). TALLY $= \{A \in \Sigma^* \mid A \subseteq \{1\}^*\}$.

Theorem 1.13 *Let $A \in$ TALLY. A is P-selective if and only if A is an initial segment of a polynomial-time computable linear ordering of $\{1\}^*$.*

We do not prove Theorem 1.13.

Definition 1.14

1. *A relation R is a* preorder *if it is reflexive and transitive.*
2. *A preorder R on Σ^* is* partially polynomial-time computable *if there is a polynomial-time computable function f such that, for all $x, y \in \Sigma^*$,*
   a) if xRy and not yRx, then $f(x, y) = f(y, x) = x$,
   b) if xRy and yRx, then $f(x, y) = f(y, x) \in \{x, y\}$, and
   c) if neither xRy nor yRx, then $f(x, y) = \#$, where $\#$ is a special symbol not in Σ.
3. *Let R be a preorder on Σ^*. We use xS_Ry as a shorthand for $xRy \wedge yRx$. S_R is an equivalence relation on Σ^*, whose set of equivalence classes is denoted by Σ^*/S_R. The relation R' on Σ^*/S_R, which is a partial ordering on Σ^*/S_R, is defined by $cl(x)R'cl(y) \iff xRy$, where $cl(z)$ denotes the member of Σ^*/S_R associated with z. S_R and R' are respectively called* the equivalence relation induced by R *and the* partial ordering induced by R.

Theorem 1.15 *A is P-selective if and only if there is a partially polynomial-time computable preorder R on Σ^* such that if S_R and R' are the equivalence relation and the partial ordering induced by R, then*

1. *R' is a linear ordering, and*
2. *A is the union of an initial segment of $(\Sigma^*/S_R, R')$.*

Proof Regarding the if direction, given R, A is P-selective via the P-selector function $f(x, y)$ that is x if xRy, and is y otherwise.

Regarding the only if direction, let A be any P-selective set. So it follows from Theorem 1.4 that A has a P-selector function f satisfying $(\forall x, y)[f(x, y) = f(y, x)]$. We define R as follows: xRy holds if and only if there exist $n \geq 1$ and strings $w_1, \ldots, w_n$ such that $x = w_1$, $y = w_n$, and, for each $1 \leq j < n$, $f(w_j, w_{j+1}) = w_j$. R is clearly a preorder, i.e., it is reflexive and transitive. It is partially polynomial-time computable via the (polynomial-time) function $f(x, y)$ itself. Note that, for each x and y, $f(x, y) = x$ (so xRy) or $f(x, y) = y$ (so yRx), since f is a P-selector. Note in particular that it will never be the case that neither xRy nor yRx holds. A is indeed the union of an initial segment of $(\Sigma^*/S_R, R')$, since if $cl(x)R'cl(y)$ and $y \in A$, then it follows easily (from the definitions of R, R', and the fact that f is a P-selector for A) that $x \in A$. And R' is clearly a linear ordering. ❏

The relationship between the structure of exponential-time complexity classes and the nature of P-selectivity within NP—and the related issue of distinguishing reductions within NP—were the historical motivation for the study of P-selective sets. For completeness, we mention here, without proof, some key results along this line, and the Bibliographic Notes provide pointers for the reader interested in this topic.

Theorem 1.16 *For every tally language A there exist sets B and C such that*

1. $B \leq_{ptt}^p A$,
2. $A \leq_T^p B$,
3. $C \leq_{tt}^p A$,
4. $A \leq_T^p C$,
5. $B \leq_{ptt}^p C$,
6. $C \leq_{tt}^p B$,
7. B *is* P-*selective, and*
8. *if* C *is* P-*selective then* $C \in$ P.

Using the fact that if $E \neq NE$ then $NP - P$ contains tally sets, and using various downward closure properties of the P-selective sets and the P sets, the above theorem yields the following corollary.

Corollary 1.17 *If* $E \neq NE$ *then there exist sets B and C such that*

1. $B \in NP - P$,
2. $B \leq_{ptt}^p C$,
3. $C \not\leq_{ptt}^p B$, *and*
4. $\overline{B} \not\leq_{pos}^p B$.

Thus, if $E \neq NE$, *there exists an* NP $\leq_{tt}^p$-*degree[2] other than that of* P *that consists of at least two* $\leq_{pos}^p$-*degrees. In particular, if* $E \neq NE$, *then there exists an* NP $\leq_T^p$-*degree other than that of* P *that consists of at least two* $\leq_m^p$-*degrees.*

Corollary 1.18 *If* $NE \cap coNE \neq E$ *then there exist sets B and C in* $NP - P$ *such that $B \leq_{ptt}^p C$, $C \leq_{tt}^p B$, and $C \not\leq_{ptt}^p B$. Thus, in particular, B and C are (polynomial-time) Turing equivalent but are not (polynomial-time) many-one equivalent.*

Corollary 1.19 $NP \cap P\text{-sel} = P \implies NP \subseteq E = NE$.

In particular, note that there are P-selective sets that are arbitrarily complex. That is, from Theorem 1.16, for any tally set D there is a P-selective set A_D such that $D \leq_T^p A_D$. Thus, for example, there are P-selective sets that are not in $\bigcup_{k>0} DTIME[2^{2^{kn}}]$, and indeed there are P-selective sets that are not in the arithmetical hierarchy (informally put, that are horrifically far from being decidable). These facts highlight the striking way the P-selective sets straddle the requirements of simplicity and complexity. The P-selective sets are so structurally simple that (unless $P = NP$) no P-selective set can be NP-complete, yet even complexity classes vastly larger than NP fail to contain the

---

[2] For any reducibility $\leq_r$, an $\leq_r$-degree is an equivalence class with respect to $\equiv_r$, where we say $A \equiv_r B$ exactly if $A \leq_r B$ and $B \leq_r A$. A degree is said to be an NP degree if it contains some NP set. Note that not all sets in an NP degree need be NP sets.

P-selective sets. Similarly, the P-selective sets are so structurally simple that, for $k \geq 2$, accessing any P-selective set via k polynomially bounded quantifiers is no more powerful than accessing the same set joined with SAT via $k - 1$ polynomially bounded quantifiers (Chapter 3). Nonetheless, no number of unbounded quantifiers suffices to contain the P-selective sets.

1.2 Nondeterministic Selectivity

1.2.1 Background and Definitions

Just as NP is studied as a nondeterministic analog of P, so also are nondeterministic selectivity classes studied as analogs of P-sel. Such study is undertaken both to increase our understanding of selectivity and to increase our understanding of nondeterminism. For example, though on its surface nondeterministic selectivity might seem to have nothing to do with the interesting issue of whether NP has "unique solutions," as mentioned earlier we will see in Section 2.4 that this issue can be understood via the study of nondeterministic selectivity.

The following definition of selectivity broadens the essential idea behind P-selectivity to apply to arbitrary classes of functions—potentially including even partial and multivalued functions.

Definition 1.20 *Let f be any (possibly partial, possibly multivalued) function over two variables. For any strings x and y, set-$f(x,y)$ denotes $\{z \mid z$ is an output of $f(x,y)\}$.*

Definition 1.21

1. *Let $\mathcal{F}$ be a class of functions. We say a set A is $\mathcal{F}$-selective if there is an $f \in \mathcal{F}$ such that, for each x and y,*
   a) *set-$f(x,y) \subseteq \{x,y\}$, and*
   b) *if $A \cap \{x,y\} \neq \emptyset$ then $\emptyset \neq$ set-$f(x,y) \subseteq A$.*
   We say such a function f is an $\mathcal{F}$-selector for A.
2. *Let $\mathcal{F}$ be a class of functions. $\mathcal{F}$-sel denotes $\{A \mid A$ is $\mathcal{F}$-selective$\}$.*

Crucially, note that if exactly one of its two inputs is in the given $\mathcal{F}$-selective set, then the $\mathcal{F}$-selector function will choose (exactly) that one. On the other hand, the machine's options are somewhat relaxed in the case that all or none of its inputs are in the set. In the former case it may output both or either. In the latter case, it may output both, either, or neither.

It is not hard to see that P-sel = FP-sel, where FP denotes the total single-valued deterministic polynomial-time computable functions. However, for historical reasons, we will continue to usually use the term P-selective and the notation P-sel. We will use the notion of $\mathcal{F}$-selectivity to study selectivity with respect to the four standard nondeterministic function classes: NPSV,

$NPSV_t$, NPMV, and $NPMV_t$. In the definition below, for specificity we assume that functions have two arguments (since this is the case with selector functions).

Definition 1.22

1. *Each nondeterministic polynomial-time Turing machine is considered to be a function-computing machine as follows. Each path that rejects is considered to have no output. Each path that accepts is considered to output the string of characters stretching, at the moment that path accepts, from the left end of its semi-infinite worktape[3] through (but not including) the character underneath its worktape head. (Note that the functions thus computed are potentially partial and are potentially multivalued.)*
2. *$f \in$ NPMV if f is computed (in the sense of part 1 of this definition) by some nondeterministic polynomial-time Turing machine.*
3. *$f \in NPMV_t$ if f is total (i.e., for all x and y, $\|\text{set-}f(x,y)\| > 0$) and $f \in$ NPMV.*
4. *$f \in$ NPSV if f is single-valued (i.e., for all x and y, $\|\text{set-}f(x,y)\| \le 1$) and $f \in$ NPMV.*
5. *$f \in NPSV_t$ if f is total and $f \in$ NPSV.*

For any pair of (possibly partial, possibly multivalued) functions f and g, we take $(\forall x, y) [\text{set-}f(x,y) = \text{set-}g(x,y)]$ to be our definition of $f = g$. Immediately from the definitions, the following relationships hold.

Proposition 1.23 $FP \subseteq NPSV_t \subseteq NPSV \subseteq NPMV$ *and* $FP \subseteq NPSV_t \subseteq NPMV_t \subseteq NPMV$.

1.2.2 Basic Properties

In this section, we will prove some basic properties of the nondeterministically selective sets. We will eventually (see Theorem 4.14) prove that, unless the polynomial hierarchy collapses, none of our selectivity classes other than NPMV-sel can contain all NP sets.

We saw earlier, as Theorem 1.4, that any P-selective set has a symmetric P-selector function. We will say that a (possibly partial) multivalued 2-ary function f is *symmetric* exactly if, for all x and y, it holds that set-$f(x,y) = \text{set-}f(y,x)$. The proof we used for Theorem 1.4 does not work, even by analogy, for such partial-nondeterministic-selector-function-based classes as NPSV-sel and NPMV-sel. Nonetheless, for all four of our nondeterministic selector function classes, it is true that each set in the class belongs to the class even with respect to having some symmetric selector function. The proof below works for all four of our nondeterministic classes—and indeed, the proof approach would work well also as a proof of Theorem 1.4.

---

[3] If it has multiple tapes, we consider the first tape to be the one that contains this "output." We'll assume that the tapes in our model are semi-infinite, though it is easy to carry the notion over to other models.

Theorem 1.24 *Let C be any one of $\{\text{NPSV}_t\text{-sel}, \text{NPSV-sel}, \text{NPMV}_t\text{-sel}, \text{NPMV}_t\text{-sel}\}$. Then the following holds. If $A \in C$ then A belongs to C via some symmetric C-selector function, i.e., some C-selector function f satisfying*

$$(\forall x, y)\,[\text{set-}f(x, y) = \text{set-}f(y, x)].$$

Proof Let A be C-selective via C-selector function f'. Let $f(x, y) = f'(\min(x, y), \max(x, y))$. It is not hard to see that f is a symmetric C-selector function for A, i.e., it satisfies $(\forall x, y)\,[\text{set-}f(x, y) = \text{set-}f(y, x)]$. $\Box$

Regarding NPMV-selectivity, the notion clearly encompasses NP.

Proposition 1.25 NP $\subseteq$ NPMV-sel.

Proof Let L be in NP, and let N be a nondeterministic polynomial-time Turing machine accepting L. Consider the NPMV function $f(x, y)$ that nondeterministically guesses a computation path of $N(x)$ or $N(y)$ (i.e., each path of $N(x)$ and each path of $N(y)$ is guessed on some nondeterministic path of our machine), and that for each nondeterministically guessed computation path of $N(x)$ outputs x on that path if the path is an accepting path of $N(x)$ and outputs nothing on that path otherwise, and that for each guessed computation path of $N(y)$ outputs y on that path if the path is an accepting path of $N(y)$ and outputs nothing on that path otherwise. This NPMV function is an NPMV-selector for L since, for each x and y, set-$f(x, y) = \{x, y\} \cap L$. $\Box$

Clearly, from Proposition 1.23, we have the following inclusions.

Proposition 1.26 P-sel $\subseteq$ NPSV$_t$-sel $\subseteq$ NPSV-sel $\subseteq$ NPMV-sel. P-sel $\subseteq$ NPSV$_t$-sel $\subseteq$ NPMV$_t$-sel $\subseteq$ NPMV-sel.

In addition, note that if a set and its complement are NPMV-selective, then the set is NPMV$_t$-selective.

Definition 1.27 *For any class C, let $\text{co} \cdot C$ denote $\{A \mid \overline{A} \in C\}$.*

Theorem 1.28 NPMV-sel $\cap$ co $\cdot$ NPMV-sel = NPMV$_t$-sel.

Proof The $\supseteq$ direction is immediate since NPMV$_t$-sel is closed under complementation. Regarding the $\subseteq$ direction, suppose $A \in$ NPMV-sel $\cap$ co $\cdot$ NPMV-sel. Let f be an NPMV-selector for A and let f' be an NPMV-selector for $\overline{A}$. Then g is an NPMV$_t$-selector for A, where g is implicitly defined by

1. $a = b \implies$ set-$g(a, b) = \{a\}$, and
2. $a \neq b \implies$
   a) $a \in$ set-$g(a, b) \iff (a \in \text{set-}f(a, b) \vee b \in \text{set-}f'(a, b))$, and
   b) $b \in$ set-$g(a, b) \iff (b \in \text{set-}f(a, b) \vee a \in \text{set-}f'(a, b))$.

Note that g is an $NPMV_t$-selector for A. $\Box$

As an immediately corollary of this we have that if a set and its complement are NPSV-selective, then the set is $NPMV_t$-selective. However, it is not known whether from the same hypothesis one can conclude that the set is $NPSV_t$-selective.

Corollary 1.29   NPSV-sel $\cap$ co $\cdot$ NPSV-sel $\subseteq NPMV_t$-sel.

It turns out that results about $NPSV_t$-selectivity follow naturally and easily from results about P-selectivity. The key lemma used to make this transition is the following.

Lemma 1.30 $FP^{NP \cap coNP} = NPSV_t$.

Proof $NPSV_t \subseteq FP^{NP \cap coNP}$ holds, since given any $NPSV_t$ function f, the set

$$pre(f) = \{\langle x, y \rangle \mid (\exists z \in (\Sigma^*)^{\leq q(|x|)})[yz \in \text{set-}f(x)]\}$$

is in NP $\cap$ coNP, where q is any fixed polynomial bounding the lengths of f's outputs. It is clear that $pre(f)$ is in NP. The reason $pre(f)$ is in coNP is that since f is an $NPSV_t$ function there is an NP machine that, on input $\langle x, y \rangle$, guesses and checks the correct output value of $f(x)$ (doing so by in fact guessing paths of the $NPSV_t$ machine computing f), and each path that guessed the correct output value can easily then determine whether or not that output has y as a prefix, and such paths will accept if y is not a prefix of the real output and will reject otherwise.

Clearly an FP machine with $pre(f)$ as an oracle can, via prefix search, obtain the value of $f(x)$.

$FP^{NP \cap coNP} \subseteq NPSV_t$ is immediate by direct simulation. That is, an $NPSV_t$ function can simply simulate the FP machine and, each time a query is made, can guess both the answer and a succinct certificate of the answer, and then proceed along a given path exactly if the certificate certifies the guessed answer. $\Box$

Figure 1.1 shows the relationships between our nondeterministic selectivity classes. Though P-sel $\subseteq NPSV_t$-sel, the following result suggests it is unlikely that the two classes are equal.

Theorem 1.31   P-sel $= NPSV_t$-sel *if and only if* P $=$ NP $\cap$ coNP.

Proof   Lemma 1.30 stated that $FP^{NP \cap coNP} = NPSV_t$. So if P $=$ NP$\cap$coNP, certainly P-sel $= NPSV_t$-sel. Regarding the other direction, note the following old result, which we leave as an exercise for the reader: If P $\neq$ NP $\cap$ coNP then there is a set in (NP $\cap$ coNP) $-$ P that is not P-selective. This suffices for proving that direction, in light of the fact that every set in NP$\cap$coNP is $NPSV_t$-sel. $\Box$

On the other hand, it is not impossible that all these classes collapse, as clearly the following holds.

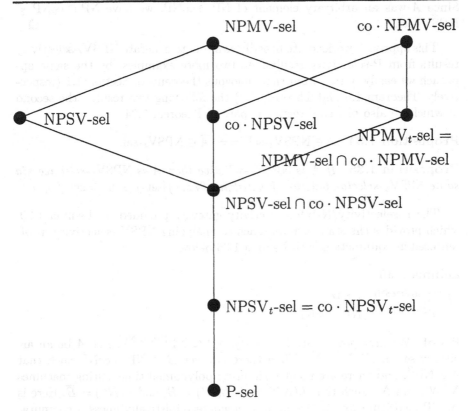

Fig. 1.1. Relationships Between Selectivity Classes

Proposition 1.32

1. $P = NP \implies$ P-sel $=$ NPSV$_t$-sel $=$ NPSV-sel $=$ NPMV$_t$-sel $=$ NPMV-sel.
2. $NP = coNP \implies$ NPSV$_t$-sel $=$ NPSV-sel $=$ NPMV$_t$-sel $=$ NPMV-sel.

Informally, the way we use Lemma 1.30 to change results about P-sel into results about NPSV$_t$-sel is to take any relativizable claim about P-sel and "relativize the entire claim with NP ∩ coNP." Of course, since NP ∩ coNP is not currently known to have complete sets, this is potentially illegal, and so one must be careful to actually argue on a "set-wise" basis. We prove the following analog of P ⊆ P-sel (Proposition 1.3) as an example.

Theorem 1.33 NP ∩ coNP ⊆ NPSV$_t$-sel.

Proof By Proposition 1.3, P ⊆ FP-sel. Proposition 1.3 clearly relativizes. Let $A \in$ NP ∩ coNP. Relativizing Proposition 1.3 by A, we have $A \in P^A \subseteq$ FP^A-sel, and by Lemma 1.30 FP^A-sel ⊆ NPSV$_t$-sel. So $A \in$ NPSV$_t$-sel.

Since A was an arbitrary element of NP $\cap$ coNP, we have NP $\cap$ coNP $\subseteq$ NPSV_t-sel. $\qquad$ $\square$

This approach works quite broadly as a tool to generate NPSV_t-selectivity results from P-selectivity results. As two more examples, by the same approach we easily obtain from the analogous theorems of Section 1.1 (respectively, Theorem 1.8 and Theorem 1.4) the following two results, the second of which we also proved directly, as part of Theorem 1.24.

Proposition 1.34 $A \in \mathrm{NPSV}_t$-sel $\iff \overline{A} \in \mathrm{NPSV}_t$-sel.

Proposition 1.35 *If A is NPSV_t-selective then A is NPSV_t-selective via some NPSV_t-selector function f satisfying $(\forall x, y)\,[\mathrm{set}\text{-}f(x, y) = \mathrm{set}\text{-}f(y, x)]$.*

The P-selectivity/NPSV_t-selectivity gateway provided by Lemma 1.30, which provides the standard approach to analyzing NPSV_t-selectivity, is often used in conjunction with Lemma 1.36 below.

Lemma 1.36

1. $\mathrm{NP}^{\mathrm{NP} \cap \mathrm{coNP}} = \mathrm{NP}$.
2. $\mathrm{P}^{\mathrm{NP} \cap \mathrm{coNP}} = \mathrm{NP} \cap \mathrm{coNP}$.

Proof We first prove part 1. Clearly, $\mathrm{NP} \subseteq \mathrm{NP}^{\mathrm{NP} \cap \mathrm{coNP}}$. Let A be an arbitrary set in $\mathrm{NP}^{\mathrm{NP} \cap \mathrm{coNP}}$. Then there is a set $B \in \mathrm{NP} \cap \mathrm{coNP}$ such that $A \in \mathrm{NP}^B$, and there are nondeterministic polynomial-time Turing machines N, N', and N'' such that $L(N^B) = A$, $L(N') = B$, and $L(N'') = \overline{B}$. Here is an NP algorithm for A. On input x, nondeterministically guess a computation path ρ of $N(x)$, guessing also the answer to each oracle query along the path. Also, for each query q that is guessed by the current path to be in B guess a computation path of $N'(q)$, and for each query q that is guessed by the current path to be in $\overline{B}$ guess a computation path of $N''(q)$. A given path accepts only if ρ is an accepting path and for every query q guessed to be in B the accompanying guessed path of $N'(q)$ is an accepting path of $N'(q)$ and for every query q guessed to be in $\overline{B}$ the accompanying guessed path of $N''(q)$ is an accepting path of $N''(q)$.

We now prove part 2. Clearly, $\mathrm{NP} \cap \mathrm{coNP} \subseteq \mathrm{P}^{\mathrm{NP} \cap \mathrm{coNP}} \subseteq \mathrm{NP}^{\mathrm{NP} \cap \mathrm{coNP}} \cap \mathrm{coNP}^{\mathrm{NP} \cap \mathrm{coNP}}$. By part 1 of this lemma, $\mathrm{NP}^{\mathrm{NP} \cap \mathrm{coNP}} \cap \mathrm{coNP}^{\overline{\mathrm{NP} \cap \mathrm{coNP}}} = \mathrm{NP} \cap \mathrm{coNP}$. $\qquad$ $\square$

Theorem 1.37 *If there exists an NP-complete NPSV_t-selective set, then $\mathrm{NP} = \mathrm{coNP}$.*

Proof The proof of Theorem 1.6 easily can be modified to yield the fact that, for each set A, it holds that:

$$\text{If SAT is } \mathrm{FP}^A\text{-selective set, then } \mathrm{NP} \subseteq \mathrm{P}^A. \quad (\star)$$

Also, it is easy to see that if there exists an NP-complete set that is NPSV_t-selective, then SAT is NPSV_t-selective.

So, suppose there is an NP-complete set D that is NPSV_t-selective. Then SAT is NPSV_t-selective, say via some NPSV_t-selector function f. By Lemma 1.30, $f \in \text{FP}^B$ for some $B \in \text{NP} \cap \text{coNP}$. From Lemma 1.36 we have that $P^B \subseteq \text{NP} \cap \text{coNP}$. So letting the A of $(\star)$ be B, we may conclude (under our current supposition) that $\text{NP} \subseteq \text{NP} \cap \text{coNP}$. Since $\text{NP} \subseteq \text{NP} \cap \text{coNP} \iff \text{NP} = \text{coNP}$, we are done. $\quad\square$

Propositions 1.34 and 1.35 and Theorem 1.37 not only provide explicit examples of deriving NPSV_t-selectivity results from P-selectivity results but also implicitly highlight the limitations of this approach. Namely, though results sometimes can be generalized even beyond NPSV_t-selectivity, the above type of almost mechanical generalizing will miss such broader generalizations. In fact, extending results to NPSV-selectivity, NPMV_t-selectivity, and NPMV-selectivity is far from mechanical. In some cases, the generalizations are obvious, but for some cases the issues remain open. Taking Proposition 1.34 as an example, it is easy to see that the NPMV_t-selective sets are also closed under complementation (Theorem 5.3). On the other hand, we will see (as Theorem 5.4) that the NPMV-selective sets are closed under complementation if and only if $\text{NP} = \text{coNP}$. As a contrasting example, Theorem 1.4 can be extended not just to NPSV_t-selectivity (Proposition 1.35) but also to all four types of nondeterministic selectivity, though a slightly different proof—which we have already provided as the proof of Theorem 1.24—is required to sidestep the fact that NPSV- and NPMV-selectors may not be total.

Finally, regarding Theorem 1.37, we will see in Section 4.4 that the analog of Theorem 1.37 holds for NPMV_t-selectivity. However, in light of Proposition 1.25, Theorem 1.37's analog holds for NPMV-selectivity only if $\text{PH} = \text{NP}$. It is an open question whether the analog of Theorem 1.37 holds for NPSV-selectivity. In fact, much of the most subtle and interesting work in the study of nondeterministic selectivity is devoted to extending results from P-selectivity (and essentially equivalently, via the discussion above, NPSV_t-selectivity) to NPSV-selectivity (see, e.g., Theorems 2.19 and 3.37).

1.3 Bibliographic Notes

Definition 1.1 is due to Selman [Sel79]. Jockusch [Joc68] defined the semi-recursive sets (Definition 1.11) and discussed their use in separating the power of reducibilities on recursively enumerable degrees.

The P-close sets were first studied as such by Schöning [Sch86]. The near-testable sets were defined by Goldsmith et al. [GJY87,GHJY91]. NNT, the implicitly membership-testable sets, was defined by Hemaspaandra and

Hoene [HH91]. Quasipolynomial time has been studied mostly via the construction of specific algorithms running in that time bound, and more recently this notion has been studied as a complexity class by Barrington, who introduced the notation qP ([Bar92,Bar95], see also [BI97]). Almost polynomial time (APT) was first studied by Meyer and Paterson [MP79].

Theorem 1.2 is due to Selman [Sel81] (see also [Ko83,Sel79,Ko82] regarding this result, standard left cuts, and other types of left cuts). The work, alluded to in the Preface, linking time-bounded left cuts to a refinement of semifeasible computation was done by Hemaspaandra, Zaki, and Zimand [HZZ96]. Theorem 1.4 is due to Ko [Ko83]. Theorem 1.5 and Theorem 1.6 are due to Selman [Sel79]. Self-reducibility was introduced by Schnorr [Sch76] and Meyer and Paterson [MP79], and has played a central role in complexity theory (see the survey by Joseph and Young [JY90]). Theorem 1.8 is due to Selman [Sel82b].

Definition 1.9, Definition 1.14, and Theorem 1.15 are due to Ko [Ko83]. Theorem 1.10 is due to Selman [Sel82a], as is Theorem 1.13 [Sel82b]. Proposition 1.12 is credited to McLaughlin and Appel by Jockusch [Joc68].

Theorem 1.16 and Corollaries 1.17, 1.18, and 1.19 are from the work of Selman [Sel82a,Sel79]. Selman's broad program of using P-selectivity to study the structure of polynomial-time reductions on NP appears in the papers [Sel79,Sel82b,Sel82a]. For more recent work using and extending the connection between exponential time classes and P-selectivity in NP − P, see the work of Hemaspaandra et al. [HNOS96a].

Hemaspaandra et al. [HHN$^+$95] introduced and first studied the NPSV$_t$-selective sets. The notion of selectivity via general classes of functions (Definition 1.21) is due to Hemaspaandra et al. [HNOS96b], who in particular studied the NPSV-selective sets, the NPMV-selective sets, and the NPMV$_t$-selective sets. The notion of equality for partial functions mentioned in the text just before Proposition 1.23 is the most common and natural notion of equality for partial functions and, in its single-valued case, dates back to Kleene [Kle52], who called the notion "complete equality." The distinction between complete equality and Kleene's other notion, so-called weak equality, has recently arisen, curiously enough, in the complexity-theoretic study of one-way functions ([HR99], see also [HPR01]).

The notation of Definition 1.20 is due to Selman [Sel94], and Definition 1.22 is due to Book, Long, and Selman [BLS84,BLS85]. Definition 1.27 is a standard notation in the field. Theorem 1.28 is due to Köbler [Köb95]. Lemma 1.30, Theorem 1.31, Theorem 1.33, Propositions 1.34 and 1.35, and Theorem 1.37 are due to Hemaspaandra et al. [HHN$^+$95,HHN$^+$93]. The exercise left to the reader in the proof of Theorem 1.31, namely, that if P $\neq$ NP $\cap$ coNP then (NP $\cap$ coNP) − P contains a set that is not P-selective, is a result of Selman [Sel88]. Lemma 1.36 is due to Selman [Sel79] and Schöning [Sch83] (see also [Sel74,Lon78,Sel78]).

2. Advice

2.1 Advice Strings and Circuits

As we saw in Chapter 1, no upper bound exists on the computational complexity of P-selective sets. Theorem 1.16 says that every tally language is polynomial-time Turing equivalent to some P-selective set. Since tally sets can be arbitrarily complex, so can P-selective sets.

There is another way of looking at the complexity of a set, and that is looking at the way in which information is stored in the set and how much information is stored. Let us consider the example of tally sets. Though membership testing in a tally set can be computationally very difficult for strings of the form 1^n, the question whether a string in $\Sigma^* - \{1\}^*$ belongs to the tally set is trivially answered, namely, by the answer no. Thus, tally sets have little information per length. In particular, for each fixed length n only 2 of the 2^{2^n} possible subsets of $\{0,1\}^n$ can occur in tally sets, namely $\{1^n\}$ and $\emptyset$. Thus each tally set can be completely determined by having just one bit of information for each length.

How far can we take this? Consider any set over the alphabet $\{0,1\}$. As long as the number of strings in the set for each length is small, say polynomially bounded, a similar argument holds. In the 2^n-element set $\{0,1\}^n$, every element can be described by an n-bit name (namely, its own bits). Thus for each set A for which the number of words per length is bounded by a polynomial p, there exists a polynomial q_p such that, for each n, a single string of length at most $q_p(n)$ crisply describes $A^{=n}$. In this case such a description can be, for $n > 0$, simply the concatenation of the strings of length n in A, but there are other cases—for example, certain sets of greater density—in which more subtle descriptions will be required.

The complexity of membership testing in sets, in the setting in which extra information may be used that is different for each length and such information is not charged to the complexity of the membership-testing algorithm, is part of the study known as nonuniform complexity. This type of complexity has a very attractive model in the form of families of circuits.

To recognize a tally set, all we need is one bit of information for each length. If a set has at each length only a polynomial number of elements, the set of strings of length $n > 0$ (the $n = 0$ case can also be recognized by an appropriate circuit) can be recognized by the following circuit. Our circuit starts with a layer of *and* gates, one for each element of the set. For each element b of the set, the corresponding *and* gate checks whether the input happens to be b. We ensure this by having wires to all input bits, adding *not* gates along the wires to those input bits where b is zero. These *and* gates are all input to a single *or* gate. Note that the *or* gate has output value 1 exactly when the input bits are set to one of the strings in the language. Since we'll refer to this particular structure of AND/OR/NOT circuit later, we now give it a shorthand name: a brute-force-OR-of-ANDs circuit for the given language (at a particular length).

Such a circuit can clearly be built for any subset of Σ^n, where the length of all inputs is fixed to be a single length n. But how can a subset L of Σ^* that contains different-length elements—potentially even containing elements at an infinite number of lengths—be said to be accepted via circuits? The standard approach is a natural one. Namely, we ask whether there is a family of circuits $C_0, C_1, C_2, \ldots$ such that each C_i accepts exactly $L^{=i}$.

Unfortunately, for an arbitrary language a circuit for length n of the type described above may need to consist of exponentially many gates. Thus it may not always be feasible to construct such a circuit in practice.

However, we will show in Section 2.2.1 that P-selective sets have the surprising property that a circuit family $C_0, C_1, C_2, \ldots$ having a polynomial number of gates (i.e., the number of gates in the C_i's is polynomially bounded in i) always suffices no matter how computationally complex the P-selective set is, notwithstanding the fact that the P-selective set's number of strings per length may not be bounded by any polynomial. Clearly, a collection of circuits that handle each length-n element of the set via one or more element-specific gates (for example, the brute-force-OR-of-ANDs circuit approach mentioned above yields such circuits) cannot possibly always wholly comprise the surprising, polynomial-sized circuit families for P-selective sets, since for example $\Sigma^* \in$ P-sel but a brute-force-OR-of-ANDs circuit family for Σ^* as described above would use exponential-size circuits.

Given an input string and an encoding of a circuit, it is easy to decide whether the circuit will accept the input string, i.e., whether it will output 1. The computation of the value of the circuit on the given input string can be done in time polynomial in the size of the circuit. The fact that a language A can be recognized by small circuits could be formalized by the existence of a (possibly hard to compute, or even in some cases uncomputable) appropriately output-size-bounded function h that, given as input a "length" n, outputs a circuit (more formally, an encoding of a circuit) for that length.

However, though it is based on the above intuition, the most standard way to formalize the class of small circuits does so in a way that doesn't directly involve circuits. Nonetheless, this formalization is known to generate the same class of sets that would have been defined if one had done the definition directly via circuits. In particular, the standard formulation of what it means to have small circuits for a set A is that there is a polynomial-time "advice interpreter" set B, and an "advice (generation) function" h that on each input n outputs some suitably short (see below) string, such that for each x it holds that $x \in A \iff \langle x, h(|x|)\rangle \in B$. The intuition here is that $h(n)$ could try to output an encoding of a circuit handling inputs of length n, and B could be a simulator that decodes that circuit and then applies it on input x.

The key resource we measure is the number of bits that h outputs on input $|x|$, *not* the amount of time and space needed to compute $h(|x|)$. Indeed, h in some cases may not even be computable. When the number of bits of $h(|x|)$ is polynomially bounded in $|x|$, A is said to have *polynomial* advice or, equivalently, polynomial-size circuits.

Again speaking very loosely, this is somewhat like having a circuit for a special computation and not worrying about the effort that it took to obtain the circuit, but rather worrying about only whether the circuit fits within a certain space.

This notion of length-bounded advice might seem more flexible than the notion of size-bounded circuits. However, in the case described above—where the set B is in P and the advice-size limit enforces polynomial advice—the notions are equivalent.

On the other hand, B could be allowed to be in complexity classes other than P, and the advice function could be allowed to produce more (or less) than a polynomial number of bits. These variations typically give different classes of sets than the class of sets having small circuits. Again, it is important to keep in mind that $h(|x|)$ need not be an encoding of a circuit, but rather can be any kind of information that is helpful in deciding the membership of x.

Definition 2.1

1. *Let $f : \mathbb{N} \to \mathbb{N}$ be any function. Let C be any collection of sets. Define*

$$C/f = \{A \mid (\exists B \in C)\,(\exists h : \mathbb{N} \to \Sigma^*)\,[(\forall n)\,[|h(n)| = f(n)] \\ \wedge (\forall x \in \Sigma^*)\,[x \in A \iff \langle x,\, h(|x|)\rangle \in B]]\}.$$

2. *Let $\mathcal{F}$ be any class of functions mapping from $\mathbb{N}$ to $\mathbb{N}$. Define*

$$C/\mathcal{F} = \{A \mid (\exists f \in \mathcal{F})\,[A \in C/f]\}.$$

Let poly, linear, and quadratic denote the classes of functions (from $\mathbb{N}$ to $\mathbb{N}$) for which the value of the output is respectively polynomially, linearly, and quadratically bounded in the value of the input. For a collection of sets C these function classes induce, using the above definition, the advice classes $C/$poly, $C/$linear, and $C/$quadratic.

A few papers in the literature use "$|h(n)| \le f(n)$" rather than "$|h(n)| = f(n)$" in their analog of part 1 of Definition 2.1. Such a modified definition effectively allows more than $f(n)$ bits of advice since there are now $2^{f(n)+1} - 1$

possible advice strings rather than exactly $2^{f(n)}$ possible advice strings. Due to the unioning over all polynomials, however, both approaches yield the same notion of P/poly. Since it is the notion introduced in the seminal advice paper by Karp and Lipton and since it is a precise definition capturing the notion of computation given exactly $f(n)$ bits of advice, in this book we adopt the classic, standard definition (Definition 2.1), and its requirement that $|h(n)| = f(n)$.

In order to make the notion "amount of advice" even more precise, we also define a refinement that will allow us to make extremely fine-grained measurements of how much advice a set requires. Note that in Definition 2.1 the advice string of length $f(n)$ is one of $2^{f(n)}$ possible choices. Our refinement allows advice to take on one of a number of values not limited to powers of two. In particular, we introduce the notion of "k-ary advice." Note that some of the "{"s and "}"s in this definition (in particular in $C/\{g\}$, and $C/\{\mathcal{G}\}$) are syntactic objects that are part of the definition—these are not an invocation of set notation.

Definition 2.2

1. Let $g : \mathbb{N} \to \mathbb{N}^+$. We assume that natural numbers have their standard encoding over binary strings. Let C be any collection of sets. Define

$$C/\{g\} = \{A \mid (\exists B \in C)(\exists h : \mathbb{N} \to \mathbb{N}^+)[(\forall n)[h(n) \in \{1,\ldots,g(n)\}] \text{ and } (\forall x \in \Sigma^*)[x \in A \iff \langle x, h(|x|)\rangle \in B]]\}.$$

2. Let $\mathcal{G}$ be any collection of functions from $\mathbb{N}$ to $\mathbb{N}^+$. Let C be any collection of sets. Define

$$C/\{\mathcal{G}\} = \{A \mid (\exists g \in \mathcal{G})[A \in C/\{g\}]\}.$$

Lemma 2.3   For any class C closed under composition with logspace functions, and for any $f : \mathbb{N} \to \mathbb{N}$ we have, $C/\{2^f\} = C/f$.

2.2 Advice for P-Selective Sets

If P-selective sets can be recognized with only a relatively small amount of extra information, the natural first question to ask is: How small can this amount be? In this section we will derive upper and lower bounds on the length of the advice strings, and the type of advice interpreter, needed to recognize P-selective sets.

Figures 2.1 and 2.2 summarize the upper-bound results we will prove.[4] Note in particular that P-sel $\subseteq$ P/quadratic $\cap$ NP/linear $\cap$ coNP/linear. We will also show that for no time-bounded class C is P-sel $\subseteq C/n$.

---

[4] We now describe the semantics of Figure 2.1. These comments apply also to the similar figures later in the book.

   If a dot, b, can be reached via some nondownward path from another dot, a, either directly or via a series of nondownward movements, then (associating, as

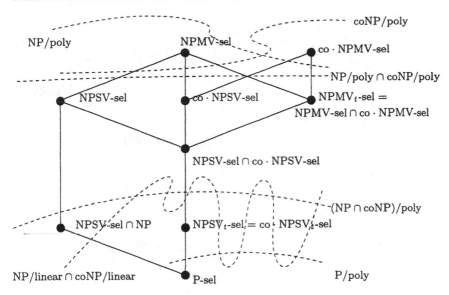

Fig. 2.1. Advice Upper Bounds for Selectivity Classes

Note: Footnote 4 describes the semantics of this figure.

2.2.1 Upper Bounds on the Amount of Advice for P-Selective Sets

Let A be a P-selective set with P-selector function f. By Theorem 1.4 we may assume without loss of generality that f is symmetric. For any pair (x, y) with $f(x, y) = y$ it holds that $x \in A \implies y \in A$. Let us write $x \leq_f y$ as a shorthand for $f(x, y) = y$. Our theorems in this section on upper bounds make use of the fact that for each P-selective set A and every finite nonempty subset B of A there exists a small collection of strings $V \subseteq B$ that is "close" to any other string in B with respect to $\leq_f$. That is, for every x in B there will be a y in V such that $y \leq_f z_1 \leq_f \cdots \leq_f z_k \leq_f x$. We can have $B = A^{=n}$, $\|V\| \leq n + 1$, and $k = 0$ (Theorem 2.4). We can also have $k = 1$ and $\|V\| = 1$ (Lemma 2.6, see also Lemma 2.21).

---

we will do throughout this footnote, items with the classes they represent) $a \subseteq b$. For example, P-sel $\subseteq$ NPMV$_t$-sel.

If a dotted line, d, passes above a dot, c, then $c \subseteq d$. For example, NPSV$_t$-sel $\subseteq$ NP/linear $\cap$ coNP/linear.

If a dotted line, g, is completely above a dotted line, f, it means that $f \subseteq g$. For example, NP/poly $\cap$ coNP/poly $\subseteq$ NP/poly.

Those are the only implications such pictures make. In particular, the fact that a dot, i, is above a dotted line, h, is not an assertion that $h \subseteq i$. For example, Figure 2.1 is not asserting that P/poly $\subseteq$ NPSV$_t$-sel; indeed, that containment happens to be untrue.

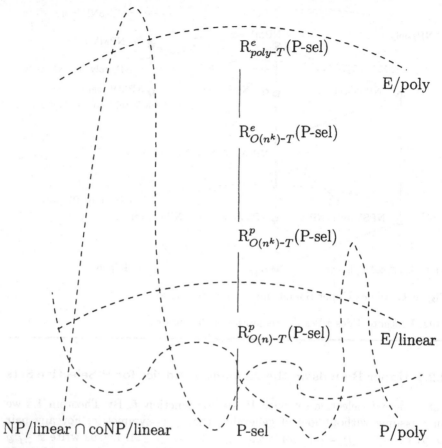

Fig. 2.2. Advice Upper Bounds for Reductions to Selectivity Classes

Note: Footnote 4 describes the semantics of this type of figure, assuming one here views P-sel and each of the "R...(P-sel)" classes of the current figure as what that footnote describes as a "dot."

Let B be any P-selective set. Let f be any symmetric P-selector function for B. Let V be all the strings at some arbitrary length n at which B is nonempty, i.e., $V = B^{=n} \neq \emptyset$. For each string y belonging to V, define $outedges(V, y) = \{(y, u) \in V \times V \mid y \neq u \wedge y \leq_f u\}$. Then $outedges(V, y) \cap outedges(V, z) = \emptyset$ for $z \neq y$ and $\|\bigcup_{y \in V} outedges(V, y)\| = \|V\|(\|V\| - 1)/2$. It follows that there is at least one y such that $\|outedges(V, y)\| \geq \lceil(\|V\| - 1)/2\rceil$. The proof of the following theorem makes use of such a counting argument. In this proof the set $Cert(V, y)$, for $y \in V$ (and thus $|y| = n$), will be the set of all (length n) strings z that y "proves" are in V via "losing" to them with respect to the P-selector function, i.e., we have $y \in V$

and $y \leq_f z$. (This indeed "proves" that $z \in V$. Note that in general it is not the case that subsets of a P-selective set respect the selector function of that P-selective set. However, since we in effect are simply looking at a single-length slice of a P-selective set, and are having contests just among strings of that length, it certainly holds that on this restricted domain, and in this case, the selector function does apply validly and informatively. In fact, in the proof we won't explicitly build the length n restriction into the *Cert* sets, but since we'll be subtracting them from sets of length-n strings, the restriction is there in effect anyway.) The argument above gives us that a $y \in V$ can be found for which the cardinality of $Cert(V, y)$ is "large," i.e., is at least $\lceil (\|V\| - 1)/2 \rceil$.

Theorem 2.4 P-sel $\subseteq$ P/poly.

Proof Let A be a P-selective set with symmetric (by Theorem 1.4) P-selector f. We will describe an advice function g such that membership of a string x in A can be determined from x and $g(|x|)$ in time polynomial in $|x| = n$. Let $V_1 = A^{=n}$. First find a string y_1 in V_1 such that $y_1 \leq_f z$ for at least $\lceil (\|V_1\| - 1)/2 \rceil$ strings $z \neq y_1$. (If $\|V_1\| = 0$ then no such string exists, which is fine.) For a set V and a string y, define $Cert(V, y) = \{u \in V \mid y \neq u \wedge y \leq_f u\}$. Remove y_1 and all the strings in $Cert(V_1, y_1)$ from V_1 and call the result V_2. In V_2 we repeat this process, i.e., if $\|V_2\| > 0$ then find a string y_2 in V_2 such that $y_2 \leq_f z$ for at least $\lceil (\|V_2\| - 1)/2 \rceil$ strings z in V_2 for which $z \neq y_2$, and remove y_2 and the strings in $Cert(V_2, y_2)$ from V_2, and call the result V_3. Continue this until some V_{k+1} satisfies $\|V_{k+1}\| = 0$. We thus create a sequence of strings $y_1, \dots, y_k$ such that for every z in $A^{=n}$ there is a y_i with $y_i \leq_f z$, and for no $z \in (\Sigma^*)^{=n} - A^{=n}$ there is a y_i with $y_i \leq_f z$.

Our logarithms in this book will all implicitly be of base two. Since the number of strings in V_{i+1} is at most half the number of strings in V_i it is easy to see that, for $\|V_1\| \geq 1$, k is at most $1 + \lceil \log \|V_1\| \rceil$. (In fact, looking a bit more closely, it is not hard to see that k is, for $\|V_1\| \geq 0$, at most $\lfloor \log(\|V_1\| + 1) \rfloor$.) Since $\|A^{=n}\| \leq 2^n$, we have $k \leq n+1$. Each of the k strings has exactly n bits. So, it is clear that there certainly exists an appropriate encoding function that encodes all our strings into an easily decodable string having exactly $n^2 + 2n + 1$ bits, namely the function that outputs $1^k 0^{n+1-k} w_1 w_2 \cdots w_{n+1}$, where $w_i = y_i$ if there is a defined y_i, and $w_i = 0^n$ otherwise. Note that $n+1$ (and thus n), and then k, and then all the w_i's, can easily be retrieved from such an advice string. The first k w_i's (and it is quite legal and possible for one of these to be 0^n) are "real" strings being coded, and the remaining w_i's, if any, are just dummy, place-filling values (that happen to each be the value 0^n). To take an extreme example, if $A^{=n} = \emptyset$, then $k = 0$ and the advice string is $0^{(n+1)^2}$. (We note in passing that for even the seemingly optimally short bit-length needed to name objects from this size universe, namely bit-length $\lceil \log(\sum_{0 \leq j \leq n+1} \binom{2^n}{j}) \rceil$, one can in P rig up a coding/decoding algorithm, though the coding schemes satisfying that will differ from the coding scheme just given.) The P language establishing that $A \in P/poly$ can now be given as follows: Given input $\langle x, a \rangle$, the P algorithm takes a string a, checks that it is of the form $1^k 0^{|x|+1-k} w_1 w_2 \cdots w_{|x|+1}$, extracts from it the strings $w_1, \ldots, w_k$, and accepts exactly if some string $y \in \{w_1, \ldots, w_k\}$ satisfies $y \leq_f x$.   ◻

Note that, though Theorem 2.4 only states inclusion of P-sel in P/poly, the proof given actually establishes P-sel $\subseteq$ P/quadratic.

A special type of P-selective set, the standard left cut, was introduced in Chapter 1. Obviously, for a standard left cut a linear amount of advice is enough: For a given length n we only need to know which is the lexicographically largest string of that length in the set, or that this set is empty at that length.

Theorem 2.5 *Let $0 \leq r < 1$. For all k, $left(r) \in P/n \ominus k$, where $a \ominus b = \max(0, a - b)$.*

Proof We first show that $left(r) \in P/n$. The only thing one has to worry about is that this might seem to require $P/\{2^n + 1\}$, namely, one token for each possible lexicographically maximum length-n string in $left(r)$, and one token for the case that $left(r)$ is empty at length n. However, the latter case occurs for more than a finite number of lengths only if $r = 0$—and trivially $left(0) \in P \subseteq P/n$. So, this case being excluded, there now are just 2^n possibilities per length. Note that we have exploited the obvious, legal flexibility of the definition to use different P sets for different r. In fact, we can, for any fixed k, code the first k bits of r into the advice interpreter. So $left(r) \in P/n \ominus k$ for each $k \geq 0$.   ◻

For classes such as $\mathcal{C}$/poly we use the phrase "advice interpreter" to describe $\mathcal{C}$, and sometimes we also use that phrase to refer to the actual $\mathcal{C}$ machine used. Using a more powerful advice interpreter than P, we can improve upon the length bound on the advice for P-selective sets. In particular, for the case of nondeterministic polynomial-time advice interpreters, linear advice suffices to recognize P-selective sets. Since the class of P-selective sets is closed under complementation, it follows that the complement of each P-selective set can be recognized with the same machinery or, equivalently, each P-selective set can be recognized by a coNP advice interpreter using linear advice.

It is possible to prove an even sharper bound (Theorem 2.7). Not only is a linear number of bits of advice sufficient to recognize a P-selective set with a nondeterministic advice interpreter, but actually $2^n + 1$ advice tokens are sufficient.

Before proving this claim in Theorem 2.7, in the following lemma we point out a nice property of P-selective sets. Any nonempty finite subset of a P-selective set has a single string, let us call it a *king* of such a subset, that has "directed distance at most two" to any other string in the subset. Curiously enough, this standard fact was first noted half a century ago in the context of animal societies. In the context of that work, the claim was: In any pride of lions, there is at least one "lion king" ℓ, i.e., a lion ℓ such that, for every lion m in the pride, either m is ℓ, or ℓ beats m, or there is some lion p such that ℓ beats p and p beats m (we assume that for any distinct pair of lions exactly one beats the other in a fight; however, beating need not be transitive).

Lemma 2.6 *Let A be a finite nonempty subset of a P-selective set B. Let f be a symmetric P-selector function for B. Then there exists a king of A (with respect to f), i.e., there is a string $x \in A$ such that*

$$(\forall y \in A)(\exists z \in A)[x \leq_f z \leq_f y].$$

Proof The proof is by simple induction on the cardinality of A. If $\|A\| = 1$, clearly the claim holds. Now, suppose $A = \{a, b\}$, $a \neq b$. If $a \leq_f b$ then we let $x = a$. Otherwise, we let $x = b$. Suppose that the lemma holds for all A' with $\|A'\| < k$ and let A have k strings. Consider any w in A. The set $A - \{w\}$ satisfies the induction hypothesis, so there exists a king $\hat{x}$ for $A - \{w\}$ such that

$$(\forall y \in A - \{w\})(\exists z \in A - \{w\})[\hat{x} \leq_f z \leq_f y].$$

If $(\exists z \in A)[\hat{x} \leq_f z \leq_f w]$ then we are done as now $\hat{x}$ is a king for all A. So suppose that $(\forall z \in A)[\hat{x} \leq_f z \implies \neg(z \leq_f w)]$. So, certainly, $(\forall z \in A)[\hat{x} \leq_f z \implies w \leq_f z]$. So $\hat{x}$ being a king for $A - \{w\}$ implies that

$$(\forall y \in A - \{w\})(\exists z \in A - \{w\})[w \leq_f z \leq_f y].$$

So clearly we have that $(\forall y \in A)(\exists z \in A)[w \leq_f z \leq_f y]$ holds and so w is a king for A. $\square$

Now we are ready to state and prove the promised theorem.

Theorem 2.7 P-sel $\subseteq$ NP$/\{2^n + 1\} \cap$ coNP$/\{2^n + 1\}$.

We can also state an upper bound in terms of length-bounded advice strings (see Definition 2.1). Then it becomes as follows, which is a slightly weaker claim than Theorem 2.7.

Corollary 2.8 P-sel $\subseteq$ NP$/n + 1 \cap$ coNP$/n + 1$.

Proof of Theorem 2.7 Let A be a P-selective set and by Theorem 1.4 let f be a symmetric P-selector for A. If $A^{=n} = \emptyset$, we use the $2^n + 1$st advice token to indicate this. Otherwise, we use Lemma 2.6 to provide a single n-bit string r_n that can be used by an NP advice interpreter to prove that an input string of length n is in A. Let us assume for length n that a string r_n with the property that $(\forall y \in A)(\exists z \in A)[r_n \leq_f z \leq_f y]$ is given as the advice string. To be more precise, given any token j other than $2^n + 1$ we view that token as indicating that the lexicographically jth string in Σ^n is

r_n. On input y of length n the NP advice interpreter can then simply guess a string z of length n and compute $f(r_n, z)$ and $f(z, y)$ to obtain a proof of membership in A for y. Of course, it is impossible to guess a convincing proof for any element of $\Sigma^n - A^{=n}$. The reason is that if $r_n \in A$ and for some z, $r_n \leq_f z \leq_f y$, then certainly $y \in A$. Finally we note that the inclusion P-sel $\subseteq$ coNP/$\{2^n + 1\}$ follows immediately from the fact that the complement of A is also a P-selective set (see Theorem 1.8), and so certainly it is in NP/$\{2^n + 1\}$ ❑   Theorem 2.7

As an immediate corollary to Theorem 2.7 we obtain that P-selective sets can be recognized in exponential time with the help of a linear amount of advice. And the exponential here is not merely $2^{n^{\mathcal{O}(1)}}$ but is even $2^{\mathcal{O}(n)}$. Why? Well, the NP advice interpreter needs to guess only one string of length n while trying to prove the membership of its input in a P-selective set. Hence, standard brute-force conversion of nondeterministic algorithms to deterministic algorithms gives a time bound on a deterministic advice interpreter of $2^{\mathcal{O}(n)}$.

Theorem 2.9   P-sel $\subseteq$ E/$\{2^n + 1\}$.

Corollary 2.10   P-sel $\subseteq$ E/linear.

The right-hand side of Theorem 2.9 is optimal, as Theorem 2.17 will show. The left-hand side is not yet quite optimal. As it turns out, we can use the above proof idea for sets that are Turing reducible to P-selective sets if we allow only a linear number of queries for the Turing reduction. Such reductions are called *linear* Turing reductions and we denote the class of sets that are reducible to P-sel via such a reduction by $R^p_{\mathcal{O}(n)\text{-}T}(\text{P-sel})$.

Theorem 2.11 $R^p_{\mathcal{O}(n)\text{-}T}(\text{P-sel}) \subseteq$ E/linear. *(Equivalently, $R^p_{\mathcal{O}(n)\text{-}T}(\text{P-sel}) \subseteq$ E/$\{2^{\mathcal{O}(n)}\}$.)*

Proof   We'll prove $R^p_{\mathcal{O}(n)\text{-}T}(\text{P-sel}) \subseteq$ E/$\{2^{\mathcal{O}(n)}\}$, which is clearly equivalent to $R^p_{\mathcal{O}(n)\text{-}T}(\text{P-sel}) \subseteq$ E/linear.

Given a (polynomial-time) linear Turing reduction machine M' reducing $A \leq^p_{\mathcal{O}(n)\text{-}T} B$, we may, and for this proof do, replace it with a machine M such that (a) there is a polynomial q and an integer j such that $M(x)$ asks at most $j|x|$ queries and runs for at most $q(|x|)$ steps regardless of what its vector of oracle answers is (even if the answers are inconsistent), and (b) $A \leq^p_{\mathcal{O}(n)\text{-}T} B$ via M. Let B be P-selective, and by Theorem 1.4 let f be a symmetric P-selector for B. On input x of length n our linear Turing reduction M will, on each branch of the tree (induced by all vectors of yes-or-no query answers) of its potential computations, query $\mathcal{O}(n)$ strings of length at most $q(n)$. Let us call the set of all of these strings $Q_M(x)$. Let $Q_M(n) = \bigcup_{|z|=n} Q_M(z)$. Note that $\|Q_M(x)\| = 2^{\mathcal{O}(n)}$ and thus $\|Q_M(n)\| = 2^{\mathcal{O}(n)}$, since there are only 2^n strings of length n. If $Q_M(n) \cap A = \emptyset$ we use the last token as advice to indicate this. Otherwise, we find a string $r_M(n)$ in $A \cap Q_M(n)$ that fits Lemma 2.6 for $A \cap Q_M(n)$ and use its index *in the set* $Q_M(n)$ as advice. Since $Q_M(n)$ is constructible by a machine simulating M on all inputs x of length n and all possible series of oracle answers, in a total time bounded by $2^{\mathcal{O}(n)}$, the string that is identified by the advice can be recovered in linear exponential time. Now after $r_M(n)$ is recovered, membership in A of any string $y \in Q_M(n)$ can be tested by successively trying every $z \in Q_M(n)$ (including y) to see whether $r_M(n) \leq_f z \leq_f y$. If $y \in A$ then there *must* be such a z for y, and if $y \in Q_M(n) - A$ there can be no such z. So acceptance or rejection of the string x by M^A can be correctly determined by a linear-exponential-time algorithm, namely via the machine that simulates M, using the just-described method of determining the answers to each oracle query. $\qquad\Box$

The reader will have noticed that in the proof above the $\mathcal{O}(n)$ limit on queries may in fact, with some adjustments in the claim, be replaced by $\mathcal{O}(n^k)$, and the reduction may, also with some adjustments in the claim, allowed to be an exponential-time machine. That is, much the same proof as just given then gives the following theorem.[5]

Theorem 2.12 *For each $k \in \mathbb{N}^+$, $R^e_{\mathcal{O}(n^k)\text{-}T}(\text{P-sel}) \subseteq E/\{2^{\mathcal{O}(n^k)}\}$.*

Corollary 2.13 $R^e_{n^{\mathcal{O}(1)}\text{-}T}(\text{P-sel}) \subseteq E/\text{poly}$.

Recall that the notation $R^e_{\mathcal{O}(n^k)\text{-}T}(\text{P-sel})$ stands for the class of sets that are reducible to a P-selective set via a reduction that uses linear exponential time and a polynomial number of queries (see Appendix A.1). In this last corollary, the inclusion also holds the other way, so we in fact have the following.

Corollary 2.14 $R^e_{n^{\mathcal{O}(1)}\text{-}T}(\text{P-sel}) = E/\text{poly}$.

Proof Let A be in E/poly. This means there is an exponential-time machine M and a function h whose outputs are of polynomially bounded length, such that $x \in A$ if and only if M accepts input $\langle x, h(|x|) \rangle$. Let $T = \{0^{\langle n,i,0 \rangle} \mid |h(n)| \leq i\} \cup \{0^{\langle n,i,1 \rangle} \mid |h(n)| \geq i$ and the ith bit of $h(n)$ is 1$\}$. By Theorem 1.16 there exists a P-selective set B such that $T \leq^p_T B$, say via polynomial-time machine M_2. Let M_1 be an exponential-time machine that has oracle B and works in two phases. On an input x working in a first phase using a simulation of M_2 to handle what in effect are queries

---

[5] One may wonder, given the proof of Theorem 2.11, why Theorem 2.12 does not have $EXP/\{2^{\mathcal{O}(n^k)}\}$ and why Corollary 2.13 does not have EXP/poly. In fact, in effect they do as $E/\{2^{\mathcal{O}(n^k)}\} = EXP/\{2^{\mathcal{O}(n^k)}\}$ and $E/\text{poly} = EXP/\text{poly}$. These equalities are subtle side effects of the definition of advice; the bits of the advice are part of the input to the machine and thus in this case may pad the input polynomially. Actually, for $E/2^{\{\mathcal{O}(n^k)\}}$ this is a bit subtle as one has to be sure to choose only large-numbered (large-sized) tokens.

to oracle T, it uses polynomially many (relative to $|x|$) queries to B to recover $h(|x|)$, which we have coded into T in such a way that it can be recovered via a polynomial number of queries to T. In the second phase, M_1 ignores its oracle and simulates $M(\langle x, h(|x|) \rangle)$. This machine in fact shows that $A \in \mathrm{R}^e_{n^{O(1)}\text{-}T}(\text{P-sel})$. $\qquad\square$

2.2.2 Are There P-Selective Sets Other Than Standard Left Cuts?

We have seen a quadratic upper bound on the amount of advice needed for P-selective sets in the case of a deterministic advice interpreter, and a linear amount of advice in the case of a nondeterministic advice interpreter. On the other hand it is obvious that standard left cuts only require a linear amount of advice even in the deterministic case. One might hope that if one could show that all P-selective sets are *very* tightly related to standard left cuts then all P-selective sets might inherit (from their related standard left cut) membership in P/linear. In this section, we will explore how closely P-selective sets are related to left cuts. We will see that if P = PP then they are closely related—but not closely enough to sustain the hypothetical attack that we just outlined.

Before our main result, we first exclude a trivial case. Though Σ^* is evidently a P-selective set, it is not a standard left cut (because of the exact definition of the notion of a standard left cut). However, if we make an exception for Σ^* then there *may* be P-selective sets that are not essentially standard left cuts, but this will be very hard to prove. Namely, a proof that there exist P-selective sets that are not many-one equivalent to standard left cuts would, by the contrapositive of the following theorem, be a proof of P $\neq$ PP. Of course, we suspect that P $\neq$ PP, but we also suspect that a proof of P $\neq$ PP is not something likely to be obtained in the next few years.

Theorem 2.15 *If* P = PP *then for every* P*-selective set* A *other than* Σ^* *there exists a standard left cut* $left(r)$ *such that* $A \equiv^p_m left(r)$.

Proof We limit ourselves to infinite P-selective sets. Finite sets are all in P and are polynomial-time many-one equivalent to some trivial standard left cut, namely $\emptyset$ satisfies $\emptyset \equiv^p_m left(0.0)$ and each nonempty set A in P satisfies $A \equiv^p_m left(0.1)$.

Consider some infinite P-selective set A with P-selector f that without loss of generality (by Theorem 1.4) we take to be a symmetric selector function. First we define some numbers. For a string y we need a name for the set of strings that it dominates, i.e., for which membership of y in A proves that such a string is also in A. So for a string y, let $dom(y) = \{z \mid |z| = |y| \wedge y \leq_f z\}$. Define $r(i) = \max\{\|dom(y)\| \mid y \in A^{=i}\}$ if $\|A^{=i}\| > 0$, and $r(i) = 0$ otherwise. For $x \in \Sigma^*$ we define for all $i \leq |x|$ the numbers $s(i, x) = \max\{\|dom(y)\| \mid x \leq_f y \wedge |y| = i\}$.

We observe that if $x \in A$ then for all $i \leq |x|$ the inequality $s(i, x) \leq r(i)$ holds. Since here $x \leq_f y$ implies that $y \in A$, it follows that $s(i, x)$ takes the maximum over a subset of the set that $r(i)$ takes the maximum over.

If $x \notin A$ then $y \in A$ implies $x \leq_f y$, so the situation is reversed and $s(i, x) \geq r(i)$ for all $i \leq |x|$. In addition, for $i = |x|$ it holds that $\|dom(x)\| \geq \|A^{=|x|} \cup \{x\}\| = \|A^{=|x|}\| + 1$ and $r(|x|) \leq \|A^{=|x|}\|$. So that in fact $s(|x|, x) > r(|x|)$ in this case.

Since for all i it holds that $r(i)$ and $s(i, x)$ are both less than or equal to 2^i, we can represent $r(i)$ and $s(i, x)$ by a binary sequence of $i + 1$ bits. So, in the definitions we now present of r and s, we take each $r(i)$ and each $s(i, x)$ to be represented (via padding by leading zeros if needed) by exactly $i + 1$ bits. The infinite bitstring $r = r(1)r(2) \cdots$ clearly has the property that $s(x) = s(1, x)s(2, x) \cdots s(|x|, x)$ is lexicographically less than or equal to r if and only if $x \in A$. To get in sync with our definition of $left(r)$, a many-one reduction may produce the string $s'(x) = s(x) - 1$ (binary subtraction) to get $s'(x) < r$ if and only if $x \in A$, provided that $s(x) \geq 1$. A nasty exception is the case where $s(x) = 0 \cdots 0$. To get around this we observe that in this case $x \notin A$ implies that also $r(|x|) = r(1)r(2) \cdots r(|x|) = 0 \cdots 0$, or, equivalently, that $s(x) = 0 \cdots 0$ and $x \notin A$ together imply that $A \cap (\Sigma^*)^{\leq |x|} = \emptyset$. So in this case $x \in A$ if and only if $s(x) < r$. We conclude that $A \leq_m^p left(r)$ if we can show that the sequence $s(x)$ can be computed from x in polynomial time. We will now show that this is the case under the assumption that $P = PP$.

Since the value $s(i, x)$, $i \leq |x|$, is at most $2^{|x|}$, it can be computed by a polynomial-time oracle machine using binary search with the help of oracle $B = \{\langle x, n, m \rangle \mid (\exists z)[|z| = n \wedge x \leq_f z \wedge \|dom(z)\| > m]\}$, which is in NP^C where $C = \{\langle z, n \rangle \mid \|\{y \mid |z| = |y| \wedge z \leq_f y\}\| > n\}$. Now a nondeterministic polynomial-time machine that on input $\langle z, n \rangle$ guesses a string y of length $|z|$ and accepts if and only if $z \leq_f y$ will have at least $n + 1$ accepting paths exactly when $\langle z, n \rangle$ in C. Hence $C \in PP$. Thus, by our $P = PP$ assumption, $C \in P$, which means that $B \in NP$, which in turn means (since $P = PP$ implies $P = NP = PP$) that $B \in P$. This means that $s(i, x)$ can be computed in polynomial time.

As it turns out, the relation $left(r) \leq_m^p A$ also holds, which we will show now. In Theorem 5.12 we will show that any set that positive Turing reduces to a P-selective set in fact many-one reduces to that same set. To show that $left(r)$ many-one reduces to A, it thus suffices to show that we have a positive Turing reduction from $left(r)$ to A.

A string x is in $left(r)$ if and only if it is lexicographically less than $r(1) \cdots r(|x|)$. We show that we can, with a positive Turing reduction on input x, recover the values of $r(i)$ for $i \leq |x|$ from the oracle A and so we can correctly decide whether $x \in left(r)$. Fix $m \leq |x|$. To compute $r(m)$ we observe that either $r(m) = 0$ if A is empty at length m or $r(m) = \max\{s(m, y) \mid |y| = m\}$ if A is not empty at length m. We show that a deterministic polynomial-time algorithm can, by querying oracle A at most a polynomial number of times, determine a string y of length m such that $s(m, y) = r(m)$ or determine that A is empty at length m and therefore $r(m) = 0$.

In the rest of this proof we consider only strings of length m and let $s(y) = s(m, y)$ for all such strings y. We have already seen that $s(y)$ is computable in polynomial time under the assumption that $P = PP$, and that $s(y) \leq 2^m$. This implies that both the sets $\{(0^m, t) \mid (\exists y)[s(y) \geq t]\}$ and $\{y \mid (\exists z)[s(y) = s(z) \wedge z <_{lex} y]\}$ are sets in NP and therefore in P under the same assumption. Therefore, there is a polynomial-time algorithm that works on an $m + 1$-bit value t (padded with zeros if need be) where $0 \leq t \leq 2^m$. It first determines by binary search the minimal value $t' \geq t$ for which a y exists with $s(y) \geq t'$. Then, if it succeeds, it returns the lexicographically minimal string with this property. If no such t' exists, the algorithm simply rejects. Let $y(t)$ be this minimal string if it exists. We have seen that $s(u) \leq r(m)$ if and only if $u \in A$. It follows that if there is any u of length m with $s(u) \geq t$ in A then $y(t) \in A$. This means that another algorithm exists that first determines whether A is empty by determining $y(1)$ using the algorithm above and querying this string. Then if A is not empty at length m, it uses the first algorithm to binary search A to determine the maximal t for which $y(t) \in A$. This maximal t equals $r(m)$.

The reduction now on input x determines these values for all $i \leq |x|$. To see that the reduction is positive observe that more strings entering A can only increase these maximal values, resulting in more strings in $left(r)$. □

Theorem 2.15 in connection with the fact that every standard left cut is Turing equivalent to some tally set yields another interesting observation.

Theorem 2.16 *If* $P = PP$ *then* P-sel $\subseteq E_T^p$(TALLY).

We are left with a somewhat strange situation regarding the amount of advice sufficing for P-selective sets. We have shown that for NP advice interpreters linear advice suffices (Theorem 2.7) and that for P advice interpreters quadratic advice suffices (Theorem 2.4). It is not clear whether this difference is inherent: Can the quadratic claim can be improved to be a linear claim? Theorem 2.7 tells us that proving a superlinear lower bound for P-interpreters proves a difference between P and NP, and Theorem 2.15 tells us that showing the existence of P-selective sets that are essentially different from standard left cuts—for which linear advice clearly suffices—is probably beyond the reach of current proof techniques. However, since P/linear is not closed downwards under $\leq_m^p$ reductions, even if $P = PP$ holds Theorem 2.15 would not directly give P-sel $\subseteq$ P/linear. Nonetheless, since P-sel $\subseteq$ NP/linear and NP $\subseteq$ PP, $P = PP$ trivially implies P-sel $\subseteq$ P/linear. That is, the hypothesis that Theorem 2.15 uses is so strong that Theorem 2.15 is hopeless anyway as an attack on P-sel's advice complexity. The current state of knowledge is that it simply is not known whether P-sel $\subseteq$ P/linear.

2.2.3 Lower Bounds

In this section we will show that the number of advice bits needed to recognize P-selective sets cannot be improved below a linear number of bits, *no matter how powerful the advice interpreter is*. For any class of advice interpreters time-bounded by some recursive function, there is a P-selective set that cannot be recognized by any such interpreter having access to less than a linear number of advice bits.

In the proof of the following theorem we will use a form of P-selective set known as a *gappy left cut*. It is a variation of the left cut and has the following structure. At certain well-defined lengths the P-selective set looks like a standard left cut, and at other lengths it looks like the empty set. When presented with two strings of the same length, a P-selector for such a set can always return the lexicographically smaller of the two. When presented with two strings of different lengths, the lengths and the complexity of the set are set up in such a way that only two cases can occur: Either the P-selective set is empty at (at least) one of the two lengths, or the larger length is so much greater than the shorter length that the P-selector can test membership of the smaller string in time polynomial in the length of the larger string, and thus can easily return a string that is "no less likely to be in the set."

Theorem 2.17 *Let $f : \mathbb{N} \to \mathbb{N}$ be any recursively computable function. Then* P-sel $\not\subseteq$ DTIME$[f]/\{2^n\}$.

Proof Let $M_1, M_2, \ldots$ be an enumeration of the (partial) recursive machines. Choose a time-constructible monotone increasing recursive function $g > f$ such that, for each n, $g(n)$ steps are enough time to simulate M_i for $i < n$ on all inputs of length less than or equal to $g(n-1)$ for all $2^{g(n-1)}$ different advice tokens for running time f applied to the given inputs. Such a function always exists. Let $g^1(0) = g(0)$ and, for all $i > 1$, let $g^i(0) = g(g^{i-1}(0))$. The set A will be nonempty only at lengths of the form $g^n(0)$, $n \geq 1$.

At stage 0, let $A_0 = \emptyset$.

At stage s, $s \geq 1$, we select the "left cut subset of $\Sigma^{g^s(0)}$" of minimal cardinality that is not recognized by M_s with *any* of the tokens from the set $\{1, 2, \ldots, 2^{g^s(0)}\}$. Call this set B and set $A_s = A_{s-1} \cup B$. Here, a "left cut subset of $\Sigma^{g^s(0)}$" means a set $V \subseteq \Sigma^{g^s(0)}$ that is either $\emptyset$ or consists of all the strings of length $g^s(0)$ that are lexicographically less than or equal to some string r of length $g^s(0)$.

This ends the construction of A. It remains to show that A is a P-selective set and that A is recognized by none of the machines running in time $f(n)$ on inputs of length n using advice from $\{1, \ldots, 2^n\}$.

Since g is time-constructible we can in polynomial time decide for a given number n whether there exists an i such that $g^i(0) = n$.

Regarding the P-selectivity of A, if x and y are strings of the same length, simply output the lexicographically smaller of the two. If x and y are strings of different lengths and $|x| = g^r(0)$ and $|y| = g^s(0)$, then without loss of generality assume that $s \geq r + 1$ (otherwise exchange x and y). By assumption, $g^s(0)$ steps suffice to simulate M_r on all inputs of length $g^r(0)$ for all $2^{g^r(0)}$ different advice tokens (note that $g^r(0) > g^{r-1}(0)$ is implicit in the definition). So in time polynomial in $|y|$ we can establish the least left cut of length $|x|$ that does not coincide with any of the sets recognized by M_r in time $g(|x|)$, and this reveals to us A at length $|x|$. We can then decide the membership of x in A and we output x if $x \in A$ and we output y otherwise.

If A is recognized for some i by machine M_i that runs in time $f(n)$ on input of length n with advice chosen from the set $\{1, \ldots, 2^n\}$, then at length $g^i(0)$ the set A coincides with some left cut that is not recognized by M_i at this length with any advice token. (Note that there are $2^{g^i(0)} + 1$ different left cuts possible at this length.) We have a contradiction. $\square$

2.3 Advice for Nondeterministically Selective Sets

In this section we will study upper bounds on the length of advice for sets that have a nondeterministic selector function. (We note in passing that nondeterministically selective sets certainly inherit the lower bounds of the P-selective sets and, in particular, inherit the bound of Theorem 2.17.)

Lemma 1.30 states that every set in $\text{NPSV}_t\text{-sel}$ has a selector function in $\text{FP}^{\text{NP} \cap \text{coNP}}$. The following theorem is an almost immediate consequence of that lemma.

Theorem 2.18 $\text{NPSV}_t\text{-sel} \subseteq (\text{NP} \cap \text{coNP})/\text{poly}$.

Proof Let A be a set in $\text{NPSV}_t\text{-sel}$. By Lemma 1.30, A has a selector function in $\text{FP}^{\text{NP} \cap \text{coNP}}$. That is, there is a set B in $\text{NP} \cap \text{coNP}$ such that A has a selector function f in FP^B. By relativization of Theorem 2.4, which indeed does relativize, A is in P^B/poly. However, $\text{P}^B \subseteq \text{P}^{\text{NP} \cap \text{coNP}} = \text{NP} \cap \text{coNP}$, so $A \in (\text{NP} \cap \text{coNP})/\text{poly}$. $\square$

NPSV-selective sets do not necessarily have total selector functions. However, for the intersection of NPSV-sel and NP we can prove a similar theorem.

Theorem 2.19 $\text{NPSV-sel} \cap \text{NP} \subseteq (\text{NP} \cap \text{coNP})/\text{poly}$.

Proof Recall from Theorem 2.4 that in a P-selective set B at length n a set V_n of at most $n+1$ strings in B can be found such that for any string x in $B^{=n}$ there is a string in V_n that proves membership of x in B with the help of the (symmetric) P-selector function. Let A be an NPSV-selective set. By the definition of an NPSV-selector f, if one of the arguments x or y is in A then set-$f(x, y)$ must be nonempty and its single element must be a string in A. By Theorem 1.24, we without loss of generality assume that our NPSV-selector function is symmetric. We can now play the game of Theorem 2.4 within the set $A^{=n}$. The NPSV-selector behaves exactly the same as the P-selector of that theorem, and here we can also find for length n a set of at most $n + 1$ strings $V_n \subseteq A^{=n}$ such that for every $x \in A^{=n}$ there is a y in V_n such that set-$f(x, y) = \{x\}$. So, to obtain a correct advice string, i.e., an advice string that only has strings in A as its encoded members, the "NPSV-sel" part of the left-hand side of the theorem's statement is enough.

However, the problem we still have is that, though it follows from set-$f(x, y) = \{x\}$ and $y \in A$ that $x \in A$, and thus that in this case also polynomial advice is sufficient, the argument of Theorem 2.4 will here yield just NPSV-sel $\cap$ NP $\subseteq$ NP/poly $\cap$ coNP/poly. The reason we do not get (NP $\cap$ coNP)/poly is that "lying" advice strings keep us from having an NP $\cap$ coNP set. To get around this we will use the NP part of the theorem's NPSV-sel $\cap$ NP hypothesis to provide certificates of membership in such a way as to allow us to indeed have an NP $\cap$ coNP set.

In particular, we will construct for each length n an advice string (of polynomial length) such that for each $x \in A^{=n}$ the string will contain among the strings it encodes some string $y \in A^{=n}$ for which set-$f(x, y) = \{x\}$. On top of that "only" an NP $\cap$ coNP predicate will be needed to show set-$f(x, y) = \{x\}$. To achieve this, we add to the strings $y_1, y_2, \ldots, y_j$, $j \leq n+1$, that make up the normal advice a collection of certificates $z_1, z_2, \ldots, z_j$ such that $y_i \in A$ can be checked by a polynomial-time verification of the string $\langle y_i, z_i \rangle$. Note that as $A \in$ NP such a witness scheme exists. Let

$$B = \{\langle 0^n, y_1, y_2, \ldots, y_j, z_1, z_2, \ldots, z_j \rangle \,|\, j \leq n + 1 \,\wedge$$
$$(\forall i : 1 \leq i \leq j)[y_i \in A \text{ is witnessed by } z_i]\}.$$

Clearly, $B \in$ P. Moreover, the following set C is in NP $\cap$ coNP.

$$C = \{\langle x, w \rangle \,|\, w = \langle 0^{|x|}, y_1, y_2, \ldots, y_j, z_1, z_2, \ldots, z_j \rangle \in B$$
$$\wedge \,(\exists i : 1 \leq i \leq j)[\text{set-}f(x, y_i) = \{x\}]\}.$$

It is immediate that $C \in$ NP: First guess the i for which set-$f(x, y_i) = \{x\}$ and then guess a computation of f that verifies this. It is also not hard to see that $C \in$ coNP: The fact that $w \in B$ guarantees, for all i, that set-$f(x, y_i) \neq \emptyset$. Then for each i and any string $x \notin y_1, y_2, \ldots, y_j$ it follows that set-$f(x, y_i) \neq \{x\}$ if and only if all computation paths returning a value return y_i, so in this case $(\forall i : 1 \leq i \leq j)[\text{set-}f(x, y_i) \neq \{x\}]$ can be tested with a *single* universal quantification. In light of this and the triviality of checking whether a string belongs to $\{y_1, y_2, \ldots, y_j\}$, we have that $C \in$ coNP.

Now if we let the advice w_n for length n consist of the lexicographically first string in B such that $w_n = \langle 0^n, y_1, y_2, \ldots, y_j, z_1, z_2, \ldots, z_j \rangle$ and for all $x \in A^{=n}$ there is a y_i such that set-$f(x, y_i) = \{x\}$ (and some special string if $A^{=n} = \emptyset$), then it holds that $x \in A$ if and only if $\langle x, w_n \rangle \in C$. Note that our advice string may not take on exactly our "polynomial" number of bits—at some lengths it may use fewer. However, we use here, as we have explicitly or implicitly used in many other places, the fact that, though Definition 2.1 requires advice of an exactly specified (by the function f of Definition 2.1) length, for many classes of the form $\mathcal{C}/\text{poly}$ we get the same advice class if we merely require that the specified advice length upper-bounds the actual advice lengths. This robustness property holds for $(\text{NP} \cap \text{coNP})/\text{poly}$, so we are done.   ☐

Note that if we had simply directly implemented a proof analogous to that of Theorem 2.4, and not demanded membership in NP or introduced certificates into the advice as membership in NP allows, we would still easily obtain the following result.

Theorem 2.20 $\text{NPSV-sel} \subseteq \text{NP}/\text{poly} \cap \text{coNP}/\text{poly}$.

For the NPMV-selective sets and the NPMV_t-selective sets we can prove a variant of this theorem. We aim only for containment in NP/poly and coNP/poly, so there will be no need for membership certificates. This makes it possible to economize on the size of the advice. We use the following nondeterministic variant of Lemma 2.6 to get a single string that proves membership for other strings.

Lemma 2.21 *Let A be a finite nonempty subset of an NPMV-selective set B. Let f be any symmetric NPSV-selector for B (by Theorem 1.24, such an f must exist). Then there exists a string x in A satisfying*

$$(\forall y \in A)(\exists z \in A)[z \in \text{set-}f(x, z) \wedge y \in \text{set-}f(z, y)].$$

Proof Replace $a \leq_f b$ by $a \in \text{set-}f(a, b)$ everywhere in the proof of Lemma 2.6. Note that since x, y, and z are all in A, neither $\text{set-}f(x, z)$ nor $\text{set-}f(z, y)$ can be empty.   ☐

Now we can show the following theorems.

Theorem 2.22 $\text{NPMV-sel} \subseteq \text{NP}/\{2^n + 1\}$.

Proof Let A be NPMV-selective via, by Theorem 1.24, symmetric selector function f. If $A^{=n}$ is nonempty, the advice needed for length n is the string, x, of Lemma 2.21 for the set $A^{=n}$. We will call this string r_n. (If $A^{=n} = \emptyset$, then we will need one special token distinct from any of the 2^n tokens assigned to represent the possible values of r_n. We will not consider this easy-to-handle special case for the rest of the proof.) Now $(\forall y \in A^{=n})(\exists z \in A^{=n})$ $[z \in \text{set-}f(r_n, z) \wedge y \in \text{set-}f(z, y)]$. The advice interpreter on input y of length n, having r_n as advice, guesses a string z and guesses proofs of (computations showing)

$$z \in \text{set-}f(r_n, z) \wedge y \in \text{set-}f(z, y).$$

Again note that neither $\text{set-}f(r_n, z)$ nor $\text{set-}f(z, y)$ is empty.   ☐

The same proof shows that $\text{NPMV}_t\text{-sel} \subseteq \text{NP}/\{2^n + 1\}$. However, NPMV_t-sel is closed under complementation (see Theorem 5.3). The actual statement is thus a bit stronger.

Theorem 2.23 $\text{NPMV}_t\text{-sel} \subseteq \text{NP}/\{2^n + 1\} \cap \text{coNP}/\{2^n + 1\}$.

Corollary 2.24 $\text{NPMV}_t\text{-sel} \subseteq \text{NP}/\text{poly} \cap \text{coNP}/\text{poly}$.

2.4 Are There Unique Solutions for NP?

2.4.1 Small Circuits and the Polynomial Hierarchy

In Chapter 1 we saw that NP $\subseteq$ P-sel would imply that P = NP (see Theorem 1.6). The proof of this statement crucially exploited the disjunctive self-reducibility of SAT. In the proof of Theorem 2.26 we also make use of this key property, though in connection with an advice interpreter rather than a P-selector.

The following proposition makes it clear that the class P/poly, a focus of Theorem 2.26, can be described in many ways. In particular, P/poly reflects the power of Turing reductions to P-selective sets.

Proposition 2.25 $R_T^p(\text{P-sel}) = R_T^p(\text{SPARSE}) = R_{tt}^p(\text{TALLY}) = \text{P/poly}$.

We now state and prove the famous Karp–Lipton Theorem.

Theorem 2.26 *If* NP $\subseteq$ P/poly *then* PH $= \Sigma_2^p$.

Proof SAT $\in$ P/poly means (see also Definition 2.1) that there is a deterministic polynomial-time Turing machine M, a polynomial q, and strings $w_0, w_1, w_2, \ldots$ such that (a) for each i, $|w_i| \leq q(i)$, and (b) for every boolean formula F, M accepts $\langle F, w_{|F|} \rangle$ if and only if F is satisfiable.

For this proof, we will need to go a bit further regarding the notion of advice that we use and the properties it provides. In particular, we will assume that the w_i's can be chosen so that each string w_i works not just for all formulas of length i but indeed for all formulas *up to and including* length i. That is, we will assume that there exist a polynomial q and a deterministic polynomial-time Turing machine, which will be a (slightly stronger than usual) advice interpreter for SAT, let us call it AI_{SAT}, such that for every length ℓ there exists a string w_ℓ, $|w_\ell| \leq q(\ell)$, such that for every boolean formula F with $|F| \leq \ell$ the following holds: AI_{SAT} accepts $\langle F, w_\ell \rangle$ if and only if $F \in$ SAT. To avoid confusion with the advice notion defined in Definition 2.1, we will (as is standard) call the just-described situation the case of having strong advice, and a string that is a w_ℓ functioning in the above setting will be called a *strong* advice string (implicitly with respect to the set, the advice interpreter, and the length ℓ). Since, as is easy to see, P/poly is the same class of languages under both notions of advice, the assumption NP $\subseteq$ P/poly implies the existence of AI_{SAT} (and an appropriate polynomial q and strings $w_0, w_1, w_2, \ldots$) in even this more demanding model.

In this proof, to remind ourselves that a string has been already ascertained to be a strong advice string, we will sometimes (somewhat redundantly) refer to strong advice strings as valid strong advice strings.

Let F be a boolean formula that has at least one free variable. We will denote by F_0 the formula obtained from F by substituting 0 for the first (in some ordering) free variable in F, and we will denote by F_1 the formula obtained from F by substituting 1 for the first free variable in F. We will assume (this is a legal assumption: an encoding exists that satisfies it) that the encoding of boolean formulas is such that, for all F, it holds that $\max(|F_0|, |F_1|) \leq |F|$. (In our actual algorithm, this will ensure that if we find a valid strong advice string for some length that is greater or equal to $|F|$, then it (in concert with AI_{SAT}) will speak correctly regarding F_0 and F_1.)

Now we wish to prove that NP $\subseteq$ P/poly implies that $\Sigma_3^p \subseteq \Sigma_2^p$. So let $L = L(M_1^{L(M_2^{\text{SAT}})})$, where M_1 and M_2 are NP oracle machines, be an arbitrary language in Σ_3^p. We without loss of generality assume that M_1 and M_2 are chosen so that they each have some polynomial such that they run in at most that particular polynomial time-bound for all oracles. We will show the existence of NP oracle machines M_3 and M_4 and a language C in NP such that $L = L(M_3^{C \oplus L(M_4)})$, which proves that $L \in \Sigma_2^p$.

Fix some x and let "$x \in L$?" be the question to be answered. We will show how M_3 and M_4 can do this job. Since M_1 and M_2 are polynomial-time machines, there is some polynomial p (namely, obtained by composing the above-mentioned polynomial bounds of M_2 and M_1) such that the length of the queries that can potentially be made by M_2 to its oracle, SAT, during the computation $M_1^{L(M_2^{\text{SAT}})}(x)$ is bounded by $p(|x|)$. To decide $x \in L$, M_3 itself nondeterministically guesses a potential strong advice string for $p(|x|)$, and then M_3 via an appropriate query to its oracle C asks C whether that guessed advice string is invalid (we will explain this in more detail later). If M_3, on a given path, finds that the strong advice string is invalid, then that path halts and rejects. However, if the path finds that the string that it guessed is a valid strong advice string, then we continue on as follows. Let $w_{p(|x|)}$ be the valid strong advice string we are working with on this path. (Different paths may find different values $w_{p(|x|)}$, so we from this point on are describing just a typical successful path, with its particular value of $w_{p(|x|)}$.) Then it (M_3 on the current path) simulates a computation of M_1 on input x, but each query y made by M_1 is replaced by a query $\langle y, w_{p(|x|)} \rangle$ to $L(M_4)$ (i.e., by the query $1\langle y, w_{p(|x|)} \rangle$ to its $C \oplus L(M_4)$ oracle). Machine M_4 will operate as follows: On any input $\langle y, w \rangle$, it simulates a computation of M_2 on input y, but each query F is replaced by computing $AI_{\text{SAT}}(\langle F, w \rangle)$. Note that each time (on the current path) our M_3, acting on input x, actually calls M_4, it calls it via asking a question of the form $\langle y, w_{p(|x|)} \rangle \in L(M_4)$, where $|y| \leq p(|x|)$, i.e., it passes it an appropriate valid strong advice string as the second argument. Note that, for any query to SAT, F, that is actually asked by the simulated M_2 in a run of M_4 that is invoked during our algorithm, $AI_{\text{SAT}}(\langle F, w_{p(|x|)} \rangle)$ accepts if and only if $F \in$ SAT. So, we see that $x \in L$ if and only if x is accepted by $M_3^{C \oplus L(M_4)}$, given that on at least one path M_3 guesses, checks (using the C part of its oracle), and passes on "up" (to the $L(M_4)$ part of its oracle) a valid strong advice string for length $p(|x|)$ (and on no path does it pass up a bad advice string).

It remains to show that there indeed exists an oracle $C \in \mathrm{NP}$ with which an invalid strong advice can be recognized as such. Let ℓ be the length for which strong advice must be computed. It follows from the assumption $\mathrm{NP} \subseteq \mathrm{P/poly}$ that strong advice for length ℓ indeed exists, so an NP machine (M_3 in our proof) can nondeterministically guess a string of the appropriate length and then check it for incorrectness (via the oracle set we are about to describe). Namely, to check guessed string w as to whether it is a valid strong advice for length m, we ask the query $\langle w, 1^m \rangle$ to the NP language C defined below (i.e., M_3 asks $0\langle w, 1^m \rangle$ to its $C \oplus L(M_4)$ oracle); if $\langle w, 1^m \rangle \in C$ then the advice string is bad, and if $\langle w, 1^m \rangle \notin C$ then the advice string is good, i.e., is a valid strong advice string.

The set C will check whether AI_{SAT} with the given advice is either wrong on any (appropriate-length) "leaves" (fully instantiated formulas) or inconsistent on any (appropriate-length) "internal nodes" (i.e., if there is some appropriate-length formula F for which it is not the case that F is—according to AI_{SAT} under the given advice—satisfiable if and only if at least one of the F_0 or F_1 is—according to AI_{SAT} under the given advice—satisfiable). Crucially note that, by induction, if AI_{SAT} is right on the leaves of a self-reduction tree, and is consistent on each internal node of that tree, then it is right at the root. C is defined as follows:

$$C = \{\langle w, 1^\ell \rangle \mid$$
$$(\exists F, |F| \leq \ell)$$
$$[((F \text{ has no free variables}) \wedge \neg(F \in \mathrm{SAT} \iff AI_{\mathrm{SAT}} \text{ accepts } \langle F, w \rangle))$$
$$\vee$$
$$((F \text{ has free variables}) \wedge$$
$$\neg(AI_{\mathrm{SAT}} \text{ accepts } \langle F, w \rangle \iff$$
$$(AI_{\mathrm{SAT}} \text{ accepts } \langle F_0, w \rangle \vee AI_{\mathrm{SAT}} \text{ accepts } \langle F_1, w \rangle)))]\}.$$

$\square$

There exist several variations on Theorem 2.26, some of which we will use later on in the book. Both of the following results have proofs generally analogous to that of Theorem 2.26.

Theorem 2.27 *If* $\mathrm{PSPACE} \subseteq \mathrm{P/poly}$ *then* $\mathrm{PSPACE} \subseteq \Sigma_2^p$.

Theorem 2.28 *If* $\mathrm{EXP} \subseteq \mathrm{P/poly}$ *then* $\mathrm{EXP} \subseteq \Sigma_2^p$.

Theorem 2.26 and the above two theorems can be strengthened, by a technique beyond this book's scope (see the Bibliographic Notes), to the following.

Theorem 2.29

1. *If* $\mathrm{NP} \subseteq (\mathrm{NP} \cap \mathrm{coNP})/\mathrm{poly}$ *then* $\mathrm{PH} \subseteq \mathrm{ZPP}^{\mathrm{NP}}$.
2. *For each* $k \geq 1$ *it holds that if* $\mathrm{PSPACE} \subseteq (\Sigma_k^p \cap \Pi_k^p)/\mathrm{poly}$ *then* $\mathrm{PSPACE} = \mathrm{ZPP}^{\Sigma_k^p}$.

3. *For each $k \geq 1$ it holds that if* $\mathrm{EXP} \subseteq (\Sigma_k^p \cap \Pi_k^p)/\mathrm{poly}$ *then* $\mathrm{EXP} = \mathrm{ZPP}^{\Sigma_k^p}$.

2.4.2 NP Lacks Unique Solutions Unless the Polynomial Hierarchy Collapses

Theorem 2.19 has a very interesting consequence for an old question regarding the solutions of a satisfiable formula. Suppose we know that a formula is satisfiable; is it possible to even nondeterministically pick a single satisfying assignment? By this we mean: Is there an NPSV function g such that $(\forall F \in \overline{\mathrm{SAT}})[\mathrm{set}\text{-}g(F) = \emptyset] \wedge (\forall F \in \mathrm{SAT})[\|\mathrm{set}\text{-}g(F)\| = 1 \wedge$ the unique element of $\mathrm{set}\text{-}g(F)$ is a satisfying assignment of $F]$? It is natural to suspect that the collections of solutions of satisfiable formulas are sufficiently unstructured that the answer is no. In the context of NP-selectivity this question is most elegantly captured by the different question (which, however, can be shown equivalent in the sense that both questions must have the same answer): Does SAT have an NPSV-selector? (To briefly hint at why the questions are equivalent, we point out that if the answer to this latter question were yes, then the disjunctive self-reducibility of SAT in combination with a hypothetical NPSV-selector would allow us to single out, given as input a satisfiable formula, a satisfying assignment in a way very similar to the approach used in the proof of in Theorem 1.6, except now involving an NPSV-selector.)

Theorem 2.31 suggests that it is unlikely that SAT has an NPSV-selector. To see this we first state what in effect is a relativized version of Theorem 2.26.

Theorem 2.30 *If* $\mathrm{NP} \subseteq (\mathrm{NP} \cap \mathrm{coNP})/\mathrm{poly}$ *then* $\mathrm{PH} = \Sigma_2^p$.

Proof Essentially, this is just Theorem 2.26 relativized by $\mathrm{NP} \cap \mathrm{coNP}$. However, since $\mathrm{NP} \cap \mathrm{coNP}$ may lack complete sets, we must do the relativization on a "per set" basis. So, suppose $\mathrm{NP} \subseteq (\mathrm{NP} \cap \mathrm{coNP})/\mathrm{poly}$. Then there is a set $B \in \mathrm{NP} \cap \mathrm{coNP}$ such that $\mathrm{SAT} \in \{B\}/\mathrm{poly}$. Since $\{B\}/\mathrm{poly} \subseteq \mathrm{P}^B/\mathrm{poly}$ and (since $B \in \mathrm{NP} \cap \mathrm{coNP}$) $\mathrm{NP}^B = \mathrm{NP}$, we have $\mathrm{NP}^B \subseteq \mathrm{P}^B/\mathrm{poly}$. However, Theorem 2.26 relativizes, and relativized by B it says $\mathrm{NP}^B \subseteq \mathrm{P}^B/\mathrm{poly} \implies \mathrm{PH}^B = \Sigma_2^{p,B}$. Since $B \in \mathrm{NP} \cap \mathrm{coNP}$ we have $\mathrm{PH}^B = \mathrm{PH}$ and $\Sigma_2^{p,B} = \Sigma_2^p$. So we conclude $\mathrm{PH} = \Sigma_2^p$. □

In light of Theorem 2.30, we can immediately see that Theorem 2.19 resolves, assuming the polynomial hierarchy does not collapse to Σ_2^p, the long-open question of whether even nondeterministic machines can latch onto a *single* satisfying assignment.

Theorem 2.31 *If* SAT *is* NPSV-*selective then* $\mathrm{PH} = \Sigma_2^p$.

Proof If SAT is NPSV-selective, then it follows from Theorem 2.19 that $\mathrm{SAT} \in (\mathrm{NP} \cap \mathrm{coNP})/\mathrm{poly}$. □

In fact, in light of part 1 of Theorem 2.29, Theorem 2.30 holds with the stronger conclusion PH = ZPP^{NP} (essentially by the same proof, but drawing on Theorem 2.29 rather than Theorem 2.26), and so, similarly, we may in Theorem 2.31 conclude that PH = ZPP^{NP}.

Also, note that SAT *is* NP2V-selective, where NP2V is the analog of NPSV, except allowing up to two output values on inputs in the domain. This leads us to an alternate formulation in which Theorem 2.31 can often be found. Given an NPMV function f, we say that g is a *refinement* of f if domain(f)=domain(g) and $(\forall x \in \Sigma^*)[\text{set-}g(x) \subseteq \text{set-}f(x)]$. The alternate formulation can now be stated.

Theorem 2.32 *If every* NP2V *function has an* NPSV *refinement then* PH = ZPP^{NP}.

2.5 Bibliographic Notes

The notion of advice, Definition 2.1, is due to Karp and Lipton [KL80]. In their seminal paper, Karp and Lipton, by studying both logarithmic and polynomial advice, stressed the importance of the notion of *length of (amount of) advice.*

Hemaspaandra and Torenvliet [HT96] define advice based on the exact number of tokens (Definition 2.2). This idea was inspired by a paper of Cai and Furst ([CF91], see also [Bar89]) on bottleneck machines.

Selman [Sel82a] proved that for any tally set, a P-selective set can be found that is polynomial-time Turing equivalent to it (see Theorem 1.16). From this it follows immediately that P-selective sets can be of arbitrary complexity. Ko [Ko83] proved that P-sel $\subseteq$ P/poly (Theorem 2.4), and indeed that P-sel $\subseteq$ P/quadratic.

Hemaspaandra and Torenvliet [HT96] proved that P-sel $\subseteq$ NP/linear and that a linear amount of advice is indeed a lower bound for recognizing P-selective sets with recursive advice interpreters (Theorem 2.17). Hemaspaandra, Nasipak, and Parkins [HNP98], using an observation due to Landau [Lan53] (Lemma 2.6), noted that the amount of nondeterminism in [HT96] can be reduced to linear, thus allowing us here, through using their approach, to much simplify the original proofs of Theorem 2.9 through Corollary 2.14, results originally obtained by Burtschick and Lindner [BL97]. Hemaspaandra et al. [HNOS96a] proved the possibly close relationship between P-selective sets and standard left cuts that is established in Theorems 2.15 and 2.16.

The inclusion properties of nondeterministic advice classes presented in this chapter, including most of the results in Section 2.3, are due to Hemaspaandra et al. [HNOS96b], except Theorem 2.20 which is due to Hemaspaandra et al. [HHN+95].

Theorems 2.26, 2.27, and 2.28 appeared in the original paper by Karp and Lipton and were later strengthened to Theorem 2.29 by Köbler and Watanabe [KW98] in a paper that introduced a beautiful proof approach known as "half-hashing." Recently, this result has been strengthened even further (note: $S_2^{NP \cap coNP} \subseteq ZPP^{NP} \subseteq \Sigma_2^p$) by Cai et al. [CCHO01] to "NP $\subseteq$ (NP $\cap$ coNP)/poly $\implies$ PH $= S_2^{NP \cap coNP}$." It follows that all the hypotheses stated in this chapter as implying PH $= ZPP^{NP}$ in fact even imply PH $= S_2^{NP \cap coNP}$. And the particular hypothesis of Theorem 2.26 even yields PH $= S_2$ [Cai01].

Theorem 2.30 was obtained independently by Kämper [Käm90] and Abadi, Feigenbaum, and Kilian [AFK89], both via quite complex proofs. The simple proof given here—essentially deriving this as an immediate corollary of the Karp–Lipton Theorem (Theorem 2.26)—is due to Hemaspaandra et al. [HHN+95]. Theorems 2.31 and 2.32 are due to Hemaspaandra et al. [HNOS96b]. Naik et al. ([NRRS98], see also [Ogi96,HOW02]) have recently obtained a result formally incomparable to Theorem 2.32: For each $k \geq 1$ it holds that: If each NP$(k + 1)$V function has an NPkV refinement then PH $= \Sigma_2^p$. It remains an open question whether the PH $= \Sigma_2^p$ here can be strengthened to PH $= ZPP^{NP}$ or to PH $= S_2^{NP \cap coNP}$. However, Hemaspaandra, Ogihara, and Wechsung [HOW02] have shown that the result of Naik et al. [NRRS98] itself is a consequence of a more general lowness result, and have provided a general sufficient condition—that they conjecture to be necessary unless the polynomial hierarchy collapses—under which one can refine multivalued nondeterministic function classes.

3. Lowness

3.1 Lowness Basics

The popular cartoon character Popeye the Sailor Man becomes much stronger when he eats spinach, but does not become any stronger when he eats, for example, cotton candy. In the vocabulary of complexity theory, we would say that "cotton candy is low for Popeye, but spinach is not low for Popeye." That is, lowness theory asks which sets, when provided to a given complexity class as oracles, fail to increase the power of that complexity class. In this chapter, we will see that P-sel, NPSV_t-sel, NPSV-sel, and NPMV_t-sel are all low for certain levels of the polynomial hierarchy, though NPMV-sel probably is not.

3.1.1 Basic Lowness Theory

The basic definition for lowness is quite simple. We (informally) say that an NP set is low for a class if, when used as an oracle, it gives no additional power to the class. The most classic low classes are lowness classes defined with respect to the levels of the polynomial hierarchy, and we now provide a more formal definition.

Definition 3.1 *(The low hierarchy)*

1. *For any complexity class $\mathcal{C}$ for which relativization is defined, let*

$$\mathrm{L}_{\mathcal{C}} = \{A \in \mathrm{NP} \mid \mathcal{C}^A = \mathcal{C}\}.$$

   If a set B belongs to $\mathrm{L}_{\mathcal{C}}$ it is sometimes described as being "low for $\mathcal{C}$" or as being $\mathcal{C}$-low.
2. *The low hierarchy's levels are $\mathrm{L}_{\Sigma_0^p}$, $\mathrm{L}_{\Sigma_1^p}$, $\mathrm{L}_{\Sigma_2^p}$, etc. We use LH to denote the entire hierarchy: $\mathrm{LH} = \mathrm{L}_{\Sigma_0^p} \cup \mathrm{L}_{\Sigma_1^p} \cup \mathrm{L}_{\Sigma_2^p} \cup \mathrm{L}_{\Sigma_3^p} \cup \cdots$.*

Note that the following inclusions hold directly from the definition of lowness.

Proposition 3.2 $\mathrm{L}_{\Sigma_0^p} \subseteq \mathrm{L}_{\Sigma_1^p} \subseteq \mathrm{L}_{\Sigma_2^p} \subseteq \cdots \subseteq \mathrm{LH}.$

$\mathrm{L}_{\mathrm{P}} = \mathrm{L}_{\Sigma_0^p}$ and $\mathrm{L}_{\mathrm{NP}} = \mathrm{L}_{\Sigma_1^p}$ have particularly simple characterizations. We will prove this by using part 1 of Lemma 1.36.

Theorem 3.3

1. $L_{\Sigma_0^p} = P$.
2. $L_{\Sigma_1^p} = NP \cap coNP$.

Proof We first prove part 1. $P \subseteq L_{\Sigma_0^p}$ since (a) $P^A = P$ for $A \in P$ and (b) $P \subseteq NP$. $L_{\Sigma_0^p} \subseteq P$ since if $P^A \subseteq P$ then, since $A \in P^A$, we have $A \in P$.

We next prove part 2. $NP \cap coNP \subseteq L_{\Sigma_1^p}$ since (a) by Lemma 1.36, $NP^A = NP$ for $A \in NP \cap coNP$ and (b) $NP \cap coNP \subseteq NP$. $L_{\Sigma_1^p} \subseteq NP \cap coNP$ since if $NP^A \subseteq NP$ then since $A \in NP^A$ we have $A \in NP$ and since $\overline{A} \in NP^A$ we also have $A \in coNP$. $\square$

The low hierarchy provides a "yardstick" with which the sets in NP can be classified. We have $L_{\Sigma_0^p} = P \subseteq L_{\Sigma_1^p} = NP \cap coNP \subseteq L_{\Sigma_2^p} \subseteq L_{\Sigma_3^p} \subseteq \cdots \subseteq LH \subseteq NP$. It is natural to ask whether the low hierarchy equals NP. It seems not to, and in fact this seeming inequality provides a key use of lowness, namely, as evidence that certain sets are not NP-complete. The following definition, proposition, and theorem capture this point.

Definition 3.4 *(The high hierarchy)*

1. $H_{\Sigma_k^p} = \{A \in NP \mid \Sigma_{k+1}^p \subseteq \Sigma_k^{p,A}\}$. *We say that $H_{\Sigma_k^p}$ is the set of languages that are "high for Σ_k^p."*
2. *The high hierarchy's levels are $H_{\Sigma_0^p}, H_{\Sigma_1^p}, H_{\Sigma_2^p}$, etc. We use HH to denote the entire hierarchy:* $HH = H_{\Sigma_0^p} \cup H_{\Sigma_1^p} \cup H_{\Sigma_2^p} \cup \cdots$.

It is immediately clear from Definition 3.4 that the high classes form a natural hierarchy, and it is not hard to see that the first two levels equal well-known classes.

Proposition 3.5

1. $H_{\Sigma_0^p} \subseteq H_{\Sigma_1^p} \subseteq H_{\Sigma_2^p} \subseteq \cdots \subseteq HH \subseteq NP$.
2. $H_{\Sigma_0^p} = \{A \mid A \text{ is } \leq_T^p\text{-complete for NP}\}$.
3. $H_{\Sigma_1^p} = \{A \mid A \text{ is } \leq_T^{sn}\text{-complete for NP}\}$.

Theorem 3.6 *For each $k \geq 0$, it holds that*

$$L_{\Sigma_k^p} \cap H_{\Sigma_k^p} \neq \emptyset \implies PH = \Sigma_k^p.$$

Proof Let $A \in L_{\Sigma_k^p} \cap H_{\Sigma_k^p}$. Since $A \in L_{\Sigma_k^p}$, $\Sigma_k^{p,A} = \Sigma_k^p$. Since $A \in H_{\Sigma_k^p}$, $\Sigma_{k+1}^p \subseteq \Sigma_k^{p,A}$. Thus, $\Sigma_{k+1}^p \subseteq \Sigma_k^p$, and so $PH = \Sigma_k^p$. $\square$

From Theorem 3.6 and Proposition 3.5, it is clear that if any low set is NP-complete then the polynomial hierarchy collapses. Indeed, the high hierarchy's levels are sometimes viewed—in light of Proposition 3.5—as successively more general notions of NP-completeness. If a low set falls anywhere within this hierarchy of generalized completeness, by Theorem 3.6 the polynomial hierarchy collapses.

Though lowness indeed is sometimes used as a tool to suggest that certain sets are not NP-complete, it should be stressed that lowness is by itself an interesting and central notion of organizational simplicity. Lowness measures the number of quantifiers needed to remove a set's ability to provide useful information. Saying a class $\mathcal{C}$ of NP sets is in $L_{\Sigma_k^p}$ says that the sets of the class $\mathcal{C}$ are of such simple organization that, informally put, in the presence of k alternating quantifiers $\mathcal{C}$ yields no more useful information than does the empty set, i.e., $\Sigma_k^{p,\mathcal{C}} = \Sigma_k^{p,\emptyset}$.

Indeed, in this chapter we study lowness primarily for the insight it gives into the degree of organizational simplicity of selectivity classes, rather than for the implicit consequences about hierarchy collapses given by lowness results, which often can be obtained—sometimes in stronger form—either directly (see Chapter 4) or via advice theory (see Chapter 2).

3.1.2 Extended Lowness and Refined Lowness

The previous section presented the classes $L_{\Sigma_k^p}$. We will also be interested in a finer-grained lowness hierarchy. Definition 3.1 defined lowness for any relativizable class. Of particular interest to us will be the "intermediate" classes[6] $L_{\Theta_k^p}$, $k \geq 1$, and $L_{\Delta_k^p}$, $k \geq 1$. It follows immediately from the definitions that $L_{\Theta_1^p} = L_{\Delta_1^p} = P$. As an exercise that isn't quite immediate, one can also check that the following inclusions hold (see Figure 3.1).

Proposition 3.7 *For all $k \geq 0$, $L_{\Sigma_k^p} \subseteq L_{\Theta_{k+1}^p} \subseteq L_{\Delta_{k+1}^p} \subseteq L_{\Sigma_{k+1}^p}$.*

All the low classes discussed so far are subclasses of NP. Because P-sel is not a subclass of NP, it is natural to extend lowness theory to include languages outside of NP. The notion known as extended lowness provides just such an extension.

Though we defined all lowness classes uniformly (i.e., via Definition 3.1), for extended lowness the definitions are a bit less uniform. In particular, we

---

[6] As noted in the appendix, our model of relativization of the Θ_k^p classes is a standard one; we assume that the query bound is enforced on the associated "Σ_{k-1}^p" class relativized with the oracle. For example, $L_{\Theta_2^p} = \{A \in NP \,|\, P^{(NP^A)[\mathcal{O}(\log n)]} = P^{NP[\mathcal{O}(\log n)]}\}$, where the "$[\mathcal{O}(\log n)]$" indicates that the base machines makes at most $\mathcal{O}(\log n)$ oracle queries (to NP^A on the left-hand side, and to NP on the right-hand side). Though this definition denies the base machine unlimited direct access to A, that is not a restriction that limits us or that defines a different class than one would get from the other plausible approach to relativizing the Θ_k^p classes. In particular, note that $P^{NP^A[\mathcal{O}(\log n)]} = P^{NP^A[\mathcal{O}(\log n)],A}$, where the right-hand side indicates that the P machine may make $\mathcal{O}(\log n)$ queries to its NP oracle and has no restriction on its access to its A oracle. The analogous equality holds for all Θ_k^p, $k \geq 1$, and all Δ_k^p, $k \geq 1$.

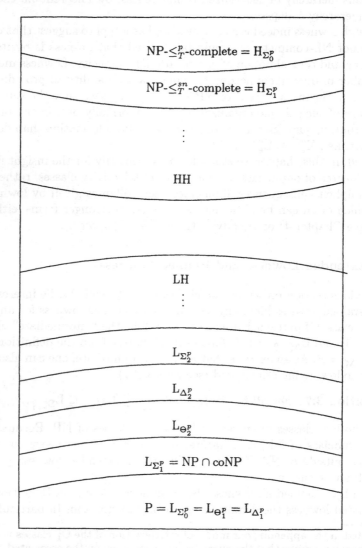

Fig. 3.1. Decomposition of NP via the Low and High Hierarchies

Note 1: This figure is unusual among our figures in its containment semantics. Here, the curved lines within the low hierarchy indicate containments. However, the curved lines within the high hierarchy indicate containments in the opposite direction than is usual in our higher-is-bigger pictures, e.g., the uppermost curved line here is intended to indicate that $H_{\Sigma_0^p} \subseteq H_{\Sigma_1^p}$.

Note 2: $LH \cap HH \neq \emptyset$ if and only if the polynomial hierarchy collapses.

start some of our definitions only at nontrivial levels, in order to avoid some anomalies at the lower levels.

Definition 3.8   *(The extended low hierarchy)*

1. *For all $k \geq 1$,*
$$\mathrm{EL}_{\Sigma_k^p} = \{A \mid \Sigma_k^{p,A} \subseteq \Sigma_{k-1}^{p,A \oplus \mathrm{SAT}}\}.$$

2. *For all $k \geq 2$,*
$$\mathrm{EL}_{\Delta_k^p} = \{A \mid \Delta_k^{p,A} \subseteq \Delta_{k-1}^{p,A \oplus \mathrm{SAT}}\}.$$

3. *For all $k \geq 2$,*
$$\mathrm{EL}_{\Theta_k^p} = \{A \mid \Theta_k^{p,A} \subseteq \Theta_{k-1}^{p,A \oplus \mathrm{SAT}}\}.$$

4. $\mathrm{ELH} = \mathrm{EL}_{\Sigma_1^p} \cup \mathrm{EL}_{\Sigma_2^p} \cup \mathrm{EL}_{\Sigma_3^p} \cup \cdots$.

The use of $A \oplus \mathrm{SAT}$ above may seem strange at first. One might expect that the extended definition might be based on asking whether A satisfies $\Sigma_k^{p,A} = \Sigma_k^p$. However, this does not generalize the notion of lowness as broadly as we intend to since the sets satisfying that equation are all members of Σ_k^p. In contrast, Definition 3.8 is broad enough to capture the intuitive flavor of lowness even in sets outside of the polynomial hierarchy.

In light of Definition 3.8, one can as an exercise verify that the following claims hold.

Proposition 3.9   *For all $k \geq 1$, $\mathrm{EL}_{\Sigma_k^p} \subseteq \mathrm{EL}_{\Theta_{k+1}^p} \subseteq \mathrm{EL}_{\Delta_{k+1}^p} \subseteq \mathrm{EL}_{\Sigma_{k+1}^p}$.*

Proposition 3.10 $\mathrm{EL}_{\Sigma_1^p} = \mathrm{EL}_{\Theta_2^p} = \mathrm{EL}_{\Delta_2^p}$.

The extended low hierarchy is closely connected to the low hierarchy by the following result, which follows because, when $A \in \mathrm{NP}$, it holds that $\Sigma_{k-1}^{p,A \oplus \mathrm{SAT}} = \Sigma_{k-1}^{p,\mathrm{SAT}} = \Sigma_k^p$, for each $k > 1$.

Proposition 3.11

1. *For all $k > 1$, $\mathrm{EL}_{\Sigma_k^p} \cap \mathrm{NP} = \mathrm{L}_{\Sigma_k^p}$. Also, $\mathrm{L}_{\Sigma_1^p} \subseteq \mathrm{EL}_{\Sigma_1^p} \cap \mathrm{NP}$.*
2. *For all $k \geq 2$, $\mathrm{EL}_{\Delta_k^p} \cap \mathrm{NP} = \mathrm{L}_{\Delta_k^p}$.*
3. *For all[7] $k \geq 3$, $\mathrm{EL}_{\Theta_k^p} \cap \mathrm{NP} = \mathrm{L}_{\Theta_k^p}$.*

---

[7] Note that we state this for $k \geq 3$. The "obvious proof" that it holds even for $k = 2$ is flawed. The reason is that, if one carefully follows the definitions, one finds that $\mathrm{EL}_{\Theta_2^p} \cap \mathrm{NP}$ is $\{A \in \mathrm{NP} \mid \mathrm{P}^{\mathrm{NP}^A[\mathcal{O}(\log n)]} \subseteq \mathrm{P}^{(\mathrm{P}^{A \oplus \mathrm{SAT}})[\mathcal{O}(\log n)]}\}$. Note that the right-hand side here is, since A is restricted to NP, equal to P^{NP}. Thus, the containment (in the context of $A \in \mathrm{NP}$) does not seem to in any obvious way imply membership of A in $\mathrm{L}_{\Theta_2^p}$, i.e., it does not seem to imply $\mathrm{P}^{\mathrm{NP}^A[\mathcal{O}(\log n)]} \subseteq \mathrm{P}^{\mathrm{NP}[\mathcal{O}(\log n)]}$. (Of course, from this discussion it is clear that if $\mathrm{P}^{\mathrm{NP}} \subseteq \mathrm{P}^{\mathrm{NP}[\mathcal{O}(\log n)]}$ then it *does* hold that $\mathrm{EL}_{\Theta_2^p} \cap \mathrm{NP} = \mathrm{L}_{\Theta_2^p}$.)

Using Proposition 3.11, we will at times obtain extended lowness results as immediate corollaries of lowness results (e.g., Corollary 3.20), and vice versa (e.g., Corollary 3.15). However, one should not assume that the low and extended low hierarchies are completely analogous. In particular, though the levels of the low hierarchy are clearly closed under many-one reductions, the following proposition that we state without proof notes the remarkable fact that the levels of the extended low hierarchy are not similarly closed.

Proposition 3.12 *For each $k \geq 2$, $EL_{\Sigma_k^p}$ is not closed under $\leq_m^p$ reductions, i.e., $R_m^p(EL_{\Sigma_k^p}) \neq EL_{\Sigma_k^p}$.*

Lowness theory and extended-lowness theory each offer two complementary challenges: the establishment of upper bounds and the establishment of lower bounds. The goal in each case is to obtain optimal results by proving adjacent upper and lower bounds. As an example, which we state without proof, it is known that the following holds.

Proposition 3.13

1. $SPARSE \subseteq EL_{\Theta_3^p}$.
2. $SPARSE \not\subseteq EL_{\Sigma_2^p}$.

However, we will see that a similarly tight result for the P-selective sets is surprisingly elusive, and is connected to a famous open question in complexity theory, namely, whether $P = PP$. Finally, we note that though one can in general hope for absolutely optimal results on extended lowness (e.g., Proposition 3.13), for lowness one currently typically seeks absolute upper bounds that in some relativized worlds have adjacent lower bounds. One is forced to adopt this weaker approach since $P = NP$ implies $P = LH$, and thus any nontrivial absolute lower bound on lowness would immediately yield a proof that $P \neq NP$.

3.2 Lowness of P-Selective Sets

In this section, we seek upper and lower bounds on the lowness and extended lowness of the class P-sel. In Section 3.2.1, we will see that all P-selective sets are in $EL_{\Sigma_2^p}$. It is an open research question whether this upper bound is tight. In particular, if $P = PP$ then P-sel $\subseteq EL_{\Sigma_1^p}$. On the other hand, in Section 3.2.2 we will note that there are relativized worlds in which there are P-selective sets that are not in $EL_{\Sigma_1^p}$ (equivalently, $EL_{\Delta_2^p}$).

Thus, under an implausible but possible structural assumption, the $EL_{\Sigma_2^p}$ bound is not optimal. On the other hand, no relativizable proof technique can strengthen the $EL_{\Sigma_2^p}$ upper bound even one "level," i.e., to an $EL_{\Delta_2^p}$ upper bound.

Though much of this section focuses on extended lowness, we also establish closely related results regarding lowness.

3.2.1 Upper Bounds

In this section, we state the (unconditional) Σ_2^p extended lowness of the P-selective sets, and we also prove that if $P = PP$ their extended lowness is even more pronounced. We will not prove Theorem 3.14 here, since we will soon—as Theorem 3.25—prove an even stronger result that easily implies Theorem 3.14.

Theorem 3.14   P-sel $\subseteq$ EL$_{\Sigma_2^p}$.

In light of Proposition 3.11, Theorem 3.14 yields the following corollary.

Corollary 3.15   P-sel $\cap$ NP $\subseteq$ L$_{\Sigma_2^p}$.

We now turn to a "conditional" extended-lowness result for the P-selective sets. We will need the following lemma.

Lemma 3.16   E$_T^p$(TALLY) $\subseteq$ EL$_{\Sigma_1^p}$.

Informally, what the following proof will do is "bring down" the appropriate tally set, and then "pass it up" to an oracle machine.

Proof of Lemma 3.16   Let $A \in \mathrm{E}_T^p(V)$, where $V \in$ TALLY. Let M_v be the deterministic polynomial-time machine such that $V = L(M_v^A)$. Let M_a be the deterministic polynomial-time machine such that $A = L(M_a^V)$, and let p_a be a polynomial bound on this machine's runtime. We must show that NP$^A \subseteq$ P$^{A \oplus \mathrm{SAT}}$. Let $B \in$ NP^A. Then $B = L(N_i^A)$ for some nondeterministic polynomial-time machine N_i for which there is a nondecreasing polynomial, q_i, that bounds the running time of N_i regardless of the oracle. We describe a P$^{A \oplus \mathrm{SAT}}$ algorithm for B.

Let x be our input. We first determine all strings in V of length up to $p_a(q_i(|x|))$, by running $M_v^A(1^m)$ for each $0 \le m \le p_a(q_i(|x|))$. Call this finite set V_0. Note that this uses queries only to A. Then, our P$^{A \oplus \mathrm{SAT}}$ machine wishes to submit the query $\langle x, V_0 \rangle$ to a certain NP set Y described below. Since SAT is part of its $A \oplus$ SAT oracle and $Y \in$ NP, our machine will in fact use any fixed many-one polynomial-time reduction from Y to SAT to convert the question for Y into a question for SAT, and then will ask the revised question to the SAT part of its oracle, and thus will get an answer that tells it whether the original query belongs to Y or not. Our machine will accept if and only if the oracle answers yes. Our NP set Y mentioned above is defined as follows:

$$Y = \{\langle x, \langle w_1, w_2, \ldots, w_j \rangle\rangle \mid j \in \mathbb{N} \wedge x \in L(N_i^{L(M_a^{\{w_1, w_2, \ldots, w_j\}})})\}.$$

This algorithm clearly accepts B, since the prefix V_0 of V is long enough to correctly determine the answers to all questions to A that can possibly be asked on the given input.

Thus $A \in$ EL$_{\Sigma_1^p}$.   ❑   Lemma 3.16

In light of Lemma 3.16, Theorem 2.16 yields the following corollary.

Corollary 3.17 *If* $P = PP$ *then* P-sel $\subseteq EL_{\Sigma_1^p}$.

Since $P = PP$ implies $P = NP$, the lowness of $NP \cap P$-sel is trivial if $P = PP$, as is reflected in the following claim.

Proposition 3.18 *If* $P = PP$ *then* $L_{\Sigma_0^p} = LH = P = NP = PH$, *and so, in particular,* P-sel $\cap NP \subseteq L_{\Sigma_0^p}$.

3.2.2 Lower Bounds

In the previous section, we noted that P-sel $\subseteq EL_{\Sigma_2^p}$, and that if $P = PP$ then P-sel $\subseteq EL_{\Sigma_1^p}$ (recall, $EL_{\Sigma_1^p} = EL_{\Delta_2^p}$). It is natural to wonder whether one can strengthen and unify these two claims, via proving that P-sel $\subseteq EL_{\Sigma_1^p}$. Unfortunately, Corollary 3.20 dims our hopes of this.

Theorem 3.19 *There is a relativized world W such that* $NP^W \cap P^W$-sel $\nsubseteq L_{\Delta_2^p}^W$.

The proof of Theorem 3.19 is quite difficult and technical, and we omit it. In light of Proposition 3.11, Theorem 3.19 immediately implies the following claim.

Corollary 3.20 *There is a relativized world W such that* P^W-sel $\nsubseteq EL_{\Delta_2^p}^W$.

Recall that "optimal" lowness results are always stated in terms of relativized optimality. This is because if $P = NP$ the low hierarchy collapses to P, and thus any nontrivial lower bound on lowness would imply $P \neq NP$—a desirable but daunting task. Thus, Corollary 3.15 and Theorem 3.19 present a typical optimal lowness claim: P-sel $\cap NP \subseteq L_{\Sigma_2^p}$ and there is a relativized world in which the $L_{\Sigma_2^p}$ bound cannot be improved one level to $L_{\Delta_2^p}$.

In striking contrast, the extended-lowness results we have for the P-selective sets are almost uniquely pathological among results on extended lowness. For almost all other classes, such as SPARSE in Proposition 3.13, the known extended-lowness results are *absolutely* optimal, as opposed to merely being optimal in some relativized world. Yet for the class P-sel we have the following claims:

1. (Theorem 3.14)   P-sel $\subseteq EL_{\Sigma_2^p}$.
2. (Corollary 3.17)   P-sel $\subseteq EL_{\Sigma_1^p}$ if $P = PP$.
3. (Corollary 3.20) P^W-sel $\nsubseteq EL_{\Delta_2^p}^W$ in some relativized world W.

That is, the extended-lowness structure of the P-selective sets, in contrast to almost all other extended-lowness results, depends on the structure of polynomial-time complexity classes. The glib explanation is that the P-selective sets, after all, are *P*-selective, and thus it is hardly surprising that a condition relating to P should potentially determine their extended lowness. However, such an explanation might leave one surprised upon learning that other classes defined in terms of P have known extended-lowness results that are optimal and unconditional. For example, the following results are well-known.

Proposition 3.21

1. $P/poly \subseteq EL_{\Theta_3^p}$.

2. $P/poly \not\subseteq EL_{\Sigma_2^p}$.

Proposition 3.22

1. P-close $\subseteq EL_{\Theta_3^p}$.

2. P-close $\not\subseteq EL_{\Sigma_2^p}$.

The "glib" explanation mentioned above still holds a measure of truth. What seemingly ties the extended-lowness structure of the P-selective sets to the power of P is not the mere fact that P appears in P-selective, but rather the specific influence of *counting* on P-selectivity. This influence is far from obvious but in fact is the substance of Theorem 2.16. The role of P is to make each P-selective set simple to describe, relative to itself, given extensive information about the number of pairs for which a certain polynomial-time test holds, namely, the number of appropriate-length strings beating a given string according to the P-selector function. P = PP makes this counting information easy to compute, and thus seems to lower the extended lowness of the P-selective sets.

We say "seems to lower" because the mere fact that there is a relativized world in which a result fails does not preclude the result holding in the real (i.e., the unrelativized) world. Thus it remains possible that one can unconditionally prove P-sel $\subseteq EL_{\Sigma_1^p}$. The existence of a relativized world in which a result fails merely says that any proof of the result is inherently a proof that does not relativize. The meaning and weight of relativized results is a controversial topic. The Bibliographic Notes section of this chapter gives pointers to some central papers on this issue.

3.3 Lowness of Nondeterministically Selective Sets

In this section, we seek upper and lower bounds on the lowness and extended lowness of the four classes of nondeterministically selective sets.

3.3.1 Upper Bounds

Figures 3.2 and 3.3 summarize the results presented in this section, i.e., the best currently known upper bounds on the extended lowness and lowness of nondeterministically selective sets. Particularly interesting is the case of the NPSV-selective sets. Though NP $\cap$ NPSV-sel $\subseteq L_{\Sigma_2^p}$, the best upper bound on NPSV-sel is NPSV-sel $\subseteq EL_{\Sigma_3^p}$. This might lead us to wonder whether NPSV-sel $\subseteq EL_{\Sigma_2^p}$. It is an open question whether this holds. On the other hand, we should mention that, though Proposition 3.11 ensures that NP$\cap\mathcal{C}$ is at least as low as $\mathcal{C}$ is extended low, it need not be the case that the converse

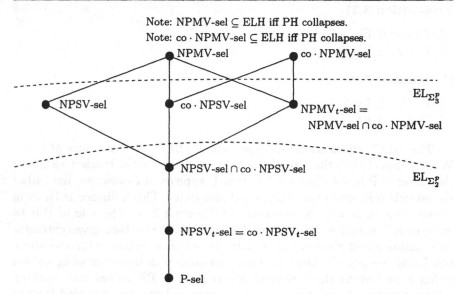

Fig. 3.2. Best Currently Known Upper Bounds on Extended Lowness of Selective Sets

Note: Footnote 4 (on page 20) describes the semantics of this type of figure.

holds. In fact, some counterexamples to the converse are known, such as those given in the following two facts.

Fact 3.23 SPARSE $\not\subseteq \mathrm{EL}_{\Sigma_2^p}$, *but* $\mathrm{NP} \cap \mathrm{SPARSE} \subseteq \mathrm{L}_{\Theta_2^p}$.

Fact 3.24 $\mathrm{R}_m^p(\mathrm{SPARSE}) \not\subseteq \mathrm{EL}_{\Sigma_2^p}$, *but* $\mathrm{NP} \cap \mathrm{R}_m^p(\mathrm{SPARSE}) \subseteq \mathrm{L}_{\Theta_2^p}$.

We now turn to proving our extended-lowness bounds.

Theorem 3.25 NPSV-sel $\cap$ co $\cdot$ NPSV-sel $\subseteq \mathrm{EL}_{\Sigma_2^p}$.

Proof Let $A \in$ NPSV-sel $\cap$ co $\cdot$ NPSV-sel. Let f_A be an NPSV-selector function for A and, via Theorem 1.24, let f_A satisfy $(\forall x, y)$ [set-$f_A(x, y) =$ set-$f_A(y, x)$]. Similarly, let $f_{\overline{A}}$ be an NPSV-selector function for $\overline{A}$ and, via Theorem 1.24, let $f_{\overline{A}}$ satisfy $(\forall x, y)$ [set-$f_{\overline{A}}(x, y) =$ set-$f_{\overline{A}}(y, x)$].

Our goal is to show that $\mathrm{NP}^{\mathrm{NP}^A} \subseteq \mathrm{NP}^{\mathrm{NP} \oplus A}$. So let B be an arbitrary set in $\mathrm{NP}^{\mathrm{NP}^A}$. We will show that $B \in \mathrm{NP}^{\mathrm{NP} \oplus A}$.

Since $B \in \mathrm{NP}^{\mathrm{NP}^A}$, there are nondeterministic polynomial-time Turing machines N_1 and N_2 such that

$$B = L(N_1^{L(N_2^A)}).$$

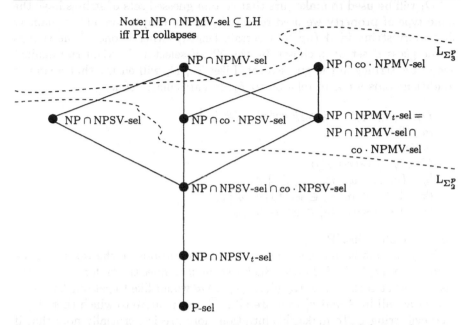

Fig. 3.3. Best Currently Known Upper Bounds on Lowness of Selective Sets

Note: Footnote 4 (on page 20) describes the semantics of this type of figure.

Let r_1 be a polynomial upper bound on the running time of N_1 and let r_2 similarly bound the running time of N_2. Without loss of generality, we may choose r_1 and r_2 so that they respectively bound the running times of N_1 and N_2 irrespective of what oracle answers the machines receive.[8]

Define the set

$$D = D_1 \oplus D_2$$

as follows.

---

[8] This issue is slightly more subtle than it might at first seem. If the N_i are drawn from a standard enumeration of machines that each themselves have a polynomial bound that holds for all oracles, then we are fine. However, if our standard enumeration has not been pre-manipulated to ensure this, then it is in fact quite possible that there are machines that for each oracle run in nondeterministic polynomial time, yet that have no single polynomial that bounds, over all oracles, their running time (see the Bibliographic Notes of this chapter for a reference). Nonetheless, even if faced here with such a machine as a potential N_1 (respectively, N_2), we can replace it with—as our real N_1 (respectively, as our real N_2)—a machine that simulates p_1 steps of this machine (respectively, p_2 steps of this machine), where p_1 is a polynomial time bound that the original N_1 happens to obey when its oracle is the specific set $L(N_2^A)$ (respectively, where p_2 is a polynomial time bound that the original N_2 happens to obey when its oracle is the specific set A).

D_1 will be used to make sure that certain guessed sets of strings have the same type of property we used repeatedly in the chapter on advice, namely that every string in A (up to a certain length) either is one of the strings in the guessed set or is chosen by the NPSV-selector f_A when run against one of the strings in the guessed set. Likewise, D_1 will ensure that a similar condition holds for $\overline{A}$ in relation to $f_{\overline{A}}$. In particular, define

$$D_1 = \{x \# v_1 \# \cdots \# v_m \# \# v'_1 \# \cdots \# v'_{m'} \mid$$
$$m \geq 0 \land$$
$$m' \geq 0 \land$$
$$(\exists y : |y| \leq r_2(r_1(|x|)))$$
$$[y \notin \{v_1, \ldots, v_m, v'_1, \ldots, v'_{m'}\} \land$$
$$(\forall j : 1 \leq j \leq m)[v_j \in \text{set-}f_A(v_j, y)] \land$$
$$(\forall j : 1 \leq j \leq m')[v'_j \in \text{set-}f_{\overline{A}}(v'_j, y)]] \}.$$

Note that D_1 is in NP.

Very informally put, the idea behind the conditions on the word y in the definition of D_1 is the following: Such a string y escapes the "information net" cast by the strings $v_1, \ldots, v_m, v'_1, \ldots, v'_{m'}$. We would like to get our hands on v_j that are all in A and v'_j that are all in $\overline{A}$, yet relative to which there is no such evil string y. To make this intuition more precise, crucially note that if we query D_1 *on an instance* $x \# v_1 \# \cdots \# v_m \# \# v'_1 \# \cdots \# v'_{m'}$ *that happens to satisfy* $\{v_1, \ldots, v_m\} \subseteq A$ *and* $\{v'_1, \ldots, v'_{m'}\} \subseteq \overline{A}$, then the D_1 oracle will say no exactly if both

1. every string of length at most $r_2(r_1(|x|))$ that is in A is either a member of $\{v_1, \ldots, v_m\}$ or is selected by the selector f_A when run against at least one member of $\{v_1, \ldots, v_m\}$, and

2. every string of length at most $r_2(r_1(|x|))$ that is in $\overline{A}$ is either a member of $\{v'_1, \ldots, v'_{m'}\}$ or is selected by the selector $f_{\overline{A}}$ when run against at least one member of $\{v'_1, \ldots, v'_{m'}\}$.

If we query D_1 on an instance that does not happen to satisfy $\{v_1, \ldots, v_m\} \subseteq A$ and $\{v'_1, \ldots, v'_{m'}\} \subseteq \overline{A}$, the claim made in the previous sentence cannot necessarily be sustained; however, in the run of the algorithm, we will query D_1 only on inputs having the desired property.

Let

$$D_2 = \{x \# v_1 \# \cdots \# v_m \# \# v'_1 \# \cdots \# v'_{m'} \mid m \geq 0 \land m' \geq 0 \land \widehat{N}_2(x \# v_1 \#$$
$$\cdots \# v_m \# \# v'_1 \# \cdots \# v'_{m'}) \text{ accepts}\},$$

where $\widehat{N}_2$ is a machine that, on any input $x \# v_1 \# \cdots \# v_m \# \# v'_1 \# \cdots \# v'_{m'}$ exactly simulates $N_2(x)$ except, whenever N_2 asks some query q, $\widehat{N}_2$ instead handles the query as follows:

1. if $q \in \{v_1, \ldots, v_m\}$ then proceed as if the query answer were yes, else
2. if $q \in \{v'_1, \ldots, v'_{m'}\}$ then proceed as if the query answer were no, else

3. nondeterministically guess a computation path ρ of (a nondeterministically guessed) one of the following list of computations:

$$f_A(v_1, q), \ldots, f_A(v_m, q), f_{\overline{A}}(v'_1, q), \ldots, f_{\overline{A}}(v'_{m'}, q).$$

Then on the guessed path ρ do the following. If the path ρ did not output q then halt and reject (on the current path). If the path ρ did output q then continue onward in the simulation of N_2, using yes as the query answer if ρ is a simulated path of f_A and using no as the query answer if ρ is a simulated path of $f_{\overline{A}}$.

Note that D_2 is in NP. Since $D = D_1 \oplus D_2$, and D_1 and D_2 are in NP, we have that $D \in$ NP.

We now show that $B \in \text{NP}^{D \oplus A}$. Consider the NP machine Q, with oracle $D \oplus A$, that on input x does the following steps.

1. **[Guess two suitably sized collections of strings, $S_A \subseteq A$ and $S_{\overline{A}} \subseteq \overline{A}$.]** First, Q nondeterministically guesses a pair of sets $S_A \subseteq (\Sigma^*)^{\le r_2(r_1(|x|))}$ and $S_{\overline{A}} \subseteq (\Sigma^*)^{\le r_2(r_1(|x|))}$ (the strange-looking composed polynomials are to account for the effect of each of the two machines potentially asking queries that are longer than their inputs) satisfying
   a) $\|S_A\| \le \log(2^{r_2(r_1(|x|))+1} - 1 + 1) = r_2(r_1(|x|)) + 1$, and
   b) $\|S_{\overline{A}}\| \le \log(2^{r_2(r_1(|x|))+1} - 1 + 1) = r_2(r_1(|x|)) + 1$.
   The equations above come from the fact that the number of strings of length at most $r_2(r_1(|x|))$ is $2^{r_2(r_1(|x|))+1} - 1$ and, essentially as noted in the proof of Theorem 2.4 except now used in the context of nondeterministic selector functions, a tournament over n' nodes has a set of the sort we want having at most $\lfloor \log(n' + 1) \rfloor$ nodes.

   Now, use our oracle A to check whether it holds that both $S_A \subseteq A$ and $S_{\overline{A}} \subseteq \overline{A}$. If either of these fails to hold, immediately reject on the current path, otherwise go on to the next step.

2. **[Check that the guessed strings have the property that for an appropriately long prefix of Σ^* it holds that each string in A either is in S_A or is chosen by the selector function f_A when run against some member of S_A, and each string in $\overline{A}$ either is in $S_{\overline{A}}$ or is chosen by the selector function $f_{\overline{A}}$ when run against some member of $S_{\overline{A}}$.]** Let $m = \|S_A\|$ and let $S_A = \{v_1, \ldots, v_m\}$. Let $m' = \|S_{\overline{A}}\|$ and let $S_{\overline{A}} = \{v'_1, \ldots, v'_{m'}\}$. Next, Q asks the question $00x\#v_1\#\cdots\#v_m\#\#v'_1\#\cdots\#v'_{m'}$ to $D \oplus A$ (i.e., it asks $x\#v_1\#\cdots\#v_m\#\#v'_1\#\cdots\#v'_{m'}$ to D_1). If the answer is yes, then the current nondeterministic path immediately halts and rejects (as the guessed set of strings does not have the desired property). Otherwise, Q proceeds to step 3. Crucially, at least one guess from step 1 will produce sets S_A and $S_{\overline{A}}$ that get the answer no from the query of the present step, since the size upper bound on our guessed sets is $r_2(r_1(|x|)) + 1$, and in

light of the standard divide and conquer argument (as in Chapter 2), this is enough to capture a set having the desired property.

3. [**Use the guessed sets to accept B without further use of A.**] Q (on its current nondeterministic path) now simulates $N_1(x)$ except that when $N_1(x)$ makes a query, q, to its oracle, Q will instead ask the query $01q\#v_1\#\cdots\#v_m\#\#v'_1\#\cdots\#v'_{m'}$ to its oracle $D \oplus A$ (i.e., it will ask the query $q\#v_1\#\cdots\#v_m\#\#v'_1\#\cdots\#v'_{m'}$ to D_2).

Note that $L(Q^{D \oplus A}) \subseteq \mathrm{NP}^{\mathrm{NP} \oplus A}$. However, by our construction, $B = L(Q^{D \oplus A})$, concluding the proof. $\square$

Corollary 3.26 $\mathrm{NPSV}_t\text{-sel} \subseteq \mathrm{EL}_{\Sigma_2^p}$.

By Proposition 3.11, this immediately yields the following corollary, which Theorem 3.37 will soon strengthen.

Corollary 3.27 $\mathrm{NP} \cap \mathrm{NPSV}_t\text{-sel} \subseteq \mathrm{L}_{\Sigma_2^p}$.

Recall from Theorem 2.20 and Corollary 2.24 that

$$\mathrm{NPSV}\text{-sel} \cup \mathrm{NPMV}_t\text{-sel} \subseteq \mathrm{NP/poly} \cap \mathrm{coNP/poly}.$$

The following lemma—which we won't prove here but which the reader may wish to prove as an exercise—allows one to translate these advice results into extended-lowness results (Theorem 3.29).

Lemma 3.28 $\mathrm{NP/poly} \cap \mathrm{coNP/poly} \subseteq \mathrm{EL}_{\Sigma_3^p}$.

Theorem 3.29

1. $\mathrm{NPSV}\text{-sel} \subseteq \mathrm{EL}_{\Sigma_3^p}$.
2. $\mathrm{NPMV}_t\text{-sel} \subseteq \mathrm{EL}_{\Sigma_3^p}$.

Proof   Immediate from Theorem 2.20, Corollary 2.24, and Lemma 3.28. $\square$

Corollary 3.30 $\mathrm{co} \cdot \mathrm{NPSV}\text{-sel} \subseteq \mathrm{EL}_{\Sigma_3^p}$.

In contrast to the case of NPMV_t-selective sets, it seems unlikely that all NPMV-selective sets are extended low.

Theorem 3.31 $\mathrm{NPMV}\text{-sel} \subseteq \mathrm{ELH}$ *only if the polynomial hierarchy collapses.*

Proof   Suppose $\mathrm{NPMV}\text{-sel} \subseteq \mathrm{ELH}$. Since by Proposition 1.25 it holds that $\mathrm{SAT} \in \mathrm{NPMV}\text{-sel}$, it follows that $\mathrm{SAT} \in \mathrm{ELH}$, and so, for some $k \geq 1$, $\Sigma_k^{p,\mathrm{SAT}} \subseteq \Sigma_{k-1}^{p,\mathrm{SAT} \oplus \mathrm{SAT}}$, i.e., $\Sigma_k^p = \Sigma_{k+1}^p$ if $k > 1$ and $\mathrm{NP}^{\mathrm{NP}} = \mathrm{P}^{\mathrm{NP}}$ if $k = 1$. So the polynomial hierarchy collapses. $\square$

Since a set A is in the extended low hierarchy if and only if $\overline{A}$ is in the extended low hierarchy, we immediately have the following corollary.

Corollary 3.32   co $\cdot$ NPMV-sel $\subseteq$ ELH *only if the polynomial hierarchy collapses.*

It is not known whether the converse of Theorem 3.31 holds. However, in light of Proposition 1.32 and Corollaries 3.17 and 3.26, we have the following partial converses.

Theorem 3.33

1. NP = coNP $\implies$ NPMV-sel $\subseteq$ EL$_{\Sigma_2^p}$.
2. P = PP $\implies$ NPMV-sel $\subseteq$ EL$_{\Sigma_1^p}$.

We have established all the extended-lowness upper bounds of Figure 3.2. We now turn to establishing the lowness upper bounds of Figure 3.3.

By Proposition 3.11, Corollary 3.30 immediately yields the following result.

Corollary 3.34   NP $\cap$ co $\cdot$ NPSV-sel $\subseteq$ L$_{\Sigma_3^p}$.

Similarly, note that since NPMV-sel $\subseteq$ NP/poly, certainly NP $\cap$ co $\cdot$ NPMV-sel $\subseteq$ (coNP/poly) $\cap$ NP $\subseteq$ NP/poly $\cap$ coNP/poly. So by Lemma 3.28 and Proposition 3.11 we even have the following.

Theorem 3.35   NP $\cap$ co $\cdot$ NPMV-sel $\subseteq$ L$_{\Sigma_3^p}$.

Corollary 3.36   NP $\cap$ NPMV$_t$-sel $\subseteq$ L$_{\Sigma_3^p}$.

Proposition 3.11 plus Theorem 3.29 imply NP $\cap$ NPSV-sel $\subseteq$ L$_{\Sigma_3^p}$. The stronger result that NP $\cap$ NPSV-sel $\subseteq$ L$_{\Theta_3^p}$ follows from Theorem 2.19 combined with an appropriate relativization of Proposition 3.21 to state, via Lemma 1.36, (NP $\cap$ coNP)/poly $\subseteq$ EL$_{\Theta_3^p}$. However, we can directly prove a still stronger lowness result for NP $\cap$ NPSV-sel by exploiting essentially the same trick used in the proof of Theorem 2.19. The proof of Theorem 3.37 is closely related to that of Theorem 3.25 in its general spirit, but has an interesting new twist in that care has to be taken to obtain certain membership proofs.

Theorem 3.37   NP $\cap$ NPSV-sel $\subseteq$ L$_{\Sigma_2^p}$.

Proof   Let $A \in$ NP $\cap$ NPSV-sel. Let f be an NPSV-selector function for A, and by Theorem 1.24 let f satisfy $(\forall x, y)$ [set-$f(x, y) =$ set-$f(y, x)$]. Since $A \in$ NP, there is a nondeterministic polynomial-time Turing machine N such that $L(N) = A$.

Our goal is to show that NP$^{\text{NP}^A}$ $\subseteq$ NP^{NP}. So let B be an arbitrary set in NP$^{\text{NP}^A}$. We will show that $B \in$ NP^{NP}.

Since $B \in$ NP$^{\text{NP}^A}$, there are nondeterministic polynomial-time Turing machines N_1 and N_2 such that $B = L(N_1^{L(N_2^A)})$. Let r_1 be a polynomial upper bound on the the running time of N_1. Let r_2 similarly bound the running time of N_2. Without loss of generality (see footnote 8 on page 51), we may choose r_1 and r_2 so that they respectively bound the running times of N_1 and N_2 irrespective of what oracle answers the machines receive. Define the set

$$D = D_1 \oplus (D_2 \oplus D_3)$$

as follows.

Let

$D_1 = \{x\#b \mid b \in \Sigma^* \text{ and some accepting computation path of } N(x) \text{ has prefix } b\}.$

Note that $D_1 \in \text{NP}$. D_1 will be used to check membership in A, and to find proofs of membership in A.

D_2 will be used to make sure that certain guessed sets of strings have the same type of property we used repeatedly in the chapter on advice, namely that they are such that every string in A (up to a certain length) either is one of the guessed strings or is chosen by the NPSV-selector f when run against one of the guessed strings. In particular, define

$D_2 = \{x\#v_1\#\cdots\#v_m \mid m \geq 0 \wedge (\exists y : |y| \leq r_2(r_1(|x|)))[y \in A \wedge y \notin \{v_1,\ldots,v_m\} \wedge (\forall j : 1 \leq j \leq m)[v_j \in \text{set-}f(v_j, y)]]\}.$

Note that the questions "$y \in A$?" and "set-$f(v_j, y)$?" can be checked by an NP machine, and in light of that we can see that $D_2 \in \text{NP}$.

Let

$D_3 = \{x\#v_1\#w_1\#\cdots\#v_m\#w_m \mid m \geq 0 \wedge (\forall j : 1 \leq j \leq m)[w_j \text{ is an accepting computation path of } N(v_j)] \wedge \widehat{N}_2^F(x\#v_1\#w_1\#\cdots\#v_m\#w_m) \text{ accepts}\},$

where $\widehat{N}_2$ is a machine that, on any input $x\#v_1\#w_1\#\cdots\#v_m\#w_m$, exactly simulates $N_2(x)$ except, whenever N_2 asks some query z, $\widehat{N}_2$ instead asks the query $z\#v_1\#w_1\#\cdots\#v_m\#w_m$ to its oracle, F.

Let $F = \{z\#v_1\#w_1\#\cdots\#v_m\#w_m \mid m \geq 0 \wedge (\forall j : 1 \leq j \leq m)[w_j \text{ is an accepting computation path of } N(v_j)] \wedge (z \in \{v_1,\ldots,v_m\} \vee (\exists j : 1 \leq j \leq m)[z \in \text{set-}f(z, v_j)])\}$. Crucially, $F \in \text{NP}\cap\text{coNP}$. This holds because F only has to run f on pairs for which it *knows* that at least one of the pair is in A— and thus for which f will be defined and single-valued. We say, informally, that it "knows" this since F, before it invokes f, can check—via the list of membership proofs $w_1,\ldots,w_m$—that $\{v_1,\ldots,v_m\} \subseteq A$.

Note that $D_3 \in \text{NP}^F$. Since $F \in \text{NP} \cap \text{coNP}$, Lemma 1.36 implies that $D_3 \in \text{NP}$. Since $D = D_1 \oplus (D_2 \oplus D_3)$ and D_1, D_2, and D_3 are in NP, we have that $D \in \text{NP}$.

We now show that $B \in \text{NP}^D$. Consider the NP machine Q, with oracle D, that on input x does the following steps.

1. **[Guess a collection of strings from A.]** First, Q nondeterministically guesses a set of at most $\log(2^{r_2(r_1(|x|))+1} - 1 + 1) = r_2(r_1(|x|)) + 1$

distinct strings of length at most $r_2(r_1(|x|))$. Then, for each string q that it guessed, it makes one query to D, namely the string $0q\#\epsilon$ (which asks $q\#\epsilon$ to D_1), in order to check that the string q is in A. If at least one of the guessed strings q is found to be in $\overline{A}$, then Q halts and rejects (on the current nondeterministic path), otherwise it goes on to step 2.]

2. **[For each guessed string find a proof that it belongs to A.]** If all the strings q guessed (in the previous step, on the current nondeterministic path) were in A (for the special case where a set of zero strings was guessed this trivially is satisfied), then Q proceeds as follows. Q uses prefix search via queries to the D_1 part of its oracle to obtain, for each string q that it guessed, a path ρ_q such that ρ_q is an accepting path of $N(q)$. Let m be the number of strings that Q guessed on the current path, and let the strings now be denoted $q_1, \ldots, q_m$.

3. **[Check that the guessed strings have the property that for a certain appropriately long prefix of A each string in A either is in the guessed set or is chosen by the selector function when run against some member of the guessed set.]** Next, Q asks the question $10x\#q_1\# \cdots \#q_m$ to D (i.e., it asks $x\#q_1\# \cdots \#q_m$ to D_2). If the answer is yes, then the current nondeterministic path immediately halts and rejects (as the guessed set of strings does not have the desired property). Otherwise, Q proceeds to step 4. Crucially, at least one guess from step 1 will produce a (possibly empty) collection of strings that are all in A and that get the answer no from the query of the present step, since the size upper bound on our guessed set is $r_2(r_1(|x|))+1$, and in light of the standard divide and conquer argument, this is enough to capture a set having the desired property (since $\log(2^{r_2(r_1(|x|))+1}-1+1) = r_2(r_1(|x|)) + 1$, see also the discussion in step 1 of the procedure given in the proof of Theorem 3.25).

4. **[Use the guessed set and the membership proofs to remove the need to use A when computing B.]** Q (on the current nondeterministic path) now simulates $N_1(x)$ except whenever $N_1(x)$ asks a query, q, to its oracle, Q will instead ask the question $11q\#q_1\#\rho_{q_1}\# \cdots \#q_m\#\rho_{q_m}$ to D (i.e., it will ask $q\#q_1\#\rho_{q_1}\# \cdots \#q_m\#\rho_{q_m}$ to D_3).

Note that $L(Q^D) \in \mathrm{NP}^D \subseteq \mathrm{NP}^{\mathrm{NP}}$. However, by our construction, $B = L(Q^D)$, concluding the proof. $\Box$

It follows immediately that $\mathrm{NP} \cap \mathrm{NPSV\text{-}sel} \cap \mathrm{co} \cdot \mathrm{NPSV\text{-}sel}$ and $\mathrm{NP} \cap \mathrm{NPSV}_t\text{-}sel$ are in $\mathrm{L}_{\Sigma_2^p}$.

Though the sets in $\mathrm{NP} \cap \mathrm{NPMV}_t\text{-}sel$ are low, it seems unlikely that the sets in $\mathrm{NP} \cap \mathrm{NPMV\text{-}sel}$ are low.

Theorem 3.38 $\mathrm{NP} \cap \mathrm{NPMV\text{-}sel} \subseteq \mathrm{LH}$ *if and only if the polynomial hierarchy collapses.*

Proof Inspection of the proof of Theorem 3.31 reveals that it already proves the "only if" direction. On the other hand, if $PH = \Sigma_k^p$ then $NP \subseteq L_{\Sigma_k^p}$ and thus $NP \cap NPMV\text{-sel} \subseteq LH$. $\square$

This completes our proof of the upper bounds stated in Figure 3.3.

3.3.2 Lower Bounds

Theorem 3.33 severely restricts our ability to state lower bounds on the extended lowness of nondeterministically selective sets. For example, proving that NPMV-sel (or $NPMV_t$-sel, NPSV-sel, or $NPSV_t$-sel) is not contained in $EL_{\Sigma_1^p}$ would immediately yield $P \neq PSPACE$, resolving one of the most important open questions of theoretical computer science. Of course, Theorem 3.19 and Corollary 3.20 implicitly provide relativized lower bounds on the lowness and extended lowness of all four types of nondeterministically selective sets. Thus, in particular, Corollary 3.27's $L_{\Sigma_2^p}$ upper bound on $NP \cap NPSV_t$-sel is optimal with respect to relativizable results.

3.4 Bibliographic Notes

Definition 3.1 is a unified statement of the lowness definitions in the literature. The $L_{\Sigma_k^p}$ classes were introduced by Schöning [Sch83] as complexity-theoretic analogs of the low and high hierarchies from recursive function theory. The $L_{\Delta_k^p}$ classes were introduced by Ko and Schöning [KS85]. The $L_{\Theta_k^p}$ classes were introduced by Long and Sheu [LS95]. Many classes have been analyzed in terms of their lowness (see the surveys by Hemaspaandra [Hem93] and Köbler [Köb95]).

Theorems 3.3 and 3.6, Definition 3.4, and Proposition 3.5 are due to Schöning [Sch83] (but Definition 3.4 adopts the now-standard redefinition of Ko and Schöning [KS85]). Regarding Definition 3.8, the $EL_{\Sigma_k^p}$ classes were introduced by Balcázar, Book, and Schöning [BBS86], the $EL_{\Delta_k^p}$ classes by Allender and Hemachandra [AH92], and the $EL_{\Theta_k^p}$ classes by Long and Sheu [LS95]. The extended low hierarchy has been shown by Sheu and Long to be an infinite hierarchy [SL94]. Proposition 3.11's three parts are due respectively to Balcázar, Book, and Schöning [BBS86], Allender and Hemachandra [AH92], and Long and Sheu [LS95].

Though the low hierarchy is very crisp and natural, the extended low hierarchy is a less intuitive notion, and exhibits some curious behaviors. For example, Hemaspaandra et al. [HJRW98] have shown that there are sets $A \notin EL_{\Sigma_2^p}$ and $B \notin EL_{\Sigma_2^p}$ such that $A \oplus B \in EL_{\Sigma_2^p}$, i.e., the join of two sets can be simpler than either of the sets. Proposition 3.12 exhibits a related, curious behavior of the extended low hierarchy. Proposition 3.12 was shown by Allender and Hemachandra [AH92] for $k = 2$, and by Vereshchagin [Ver94] for $k \geq 3$.

Lower bounds on lowness and extended lowness were introduced—and optimal lower bounds for most extended-lowness classes were obtained—by Allender and Hemachandra [AH92], who established part 2 of Proposition 3.13. Long and Sheu [LS95] proved part 1 of Proposition 3.13.

Corollary 3.15 is due to Ko and Schöning [KS85], and was generalized to Theorem 3.14 by Amir, Beigel, and Gasarch [ABG90]. Lemma 3.16 is due to Balcázar and Book [BB86]. Corollary 3.17 follows from Theorem 2.16, which is from Hemaspaandra et al. [HNOS96a].

Theorem 3.19, Corollary 3.20, part 2 of Proposition 3.21, and part 2 of Proposition 3.22 are due to Allender and Hemachandra [AH92]. Relatedly, one can certainly ask whether there are oracles relative to which the following, relativized, hold: (a) $\mathrm{NP} \cap \mathrm{co} \cdot \mathrm{NPSV}\text{-sel} \not\subseteq L_{\Sigma_2^p}$, (b) $\mathrm{NP} \cap \mathrm{NPMV}_t\text{-sel} \not\subseteq L_{\Sigma_2^p}$, (c) $\mathrm{NPSV}\text{-sel} \not\subseteq \mathrm{EL}_{\Sigma_2^p}$, or (d) $\mathrm{NPMV}_t\text{-sel} \not\subseteq \mathrm{EL}_{\Sigma_2^p}$. We commend these issues to the reader. Part 1 of Proposition 3.21 is due to Köbler [Köb94] (building on work of Gavaldà [Gav95]), and part 1 of Proposition 3.22 is due to Long and Sheu [LS95].

The meaning and weight of relativization results is a topic that has been discussed for many years. The breakthrough work of Lund et al. [LFKN92] in the development of nonrelativizable proof techniques—which led to Shamir's [Sha92] proof that the class of sets having "interactive proofs" is exactly PSPACE—has made the case for oracles more difficult (see also, e.g., [BFL91, ALM+98,HJV93,BFT98]). Nonetheless, Allender [All90] and Fortnow [For94] have eloquently argued that even in light of such techniques relativization remains an informative and useful approach (however, see also the comments of Hartmanis et al. [HCC+92]). The theory of positive relativization (see the surveys by Book [Boo89,Boo87]) links oracle results to real-world results. Partially in light of such linkages, Hemaspaandra, Ramachandran, and Zimand [HRZ95] have surveyed some long-open relativization questions whose resolution would be widely considered a major advance.

The lower bounds in Fact 3.23 and Fact 3.24 are due to Allender and Hemachandra [AH92], and the upper bounds are due to Long and Sheu [LS95]. Theorem 3.25 is due to Köbler [Köb95]. Regarding the comment within the proof of that theorem that there are machines that for each oracle run in nondeterministic polynomial time yet that have no single polynomial that bounds, over all oracles, their running time, see Cai, Hemachandra, and Vyskoč [CHV93].

Lemma 3.28 and Theorem 3.29 are from Hemaspaandra et al. [HNOS96b]. Theorems 3.31 and 3.38 follow easily from the definitions. Theorem 3.37 and Corollary 3.36 were shown by Hemaspaandra et al. [HNOS96b]. Regarding Corollary 3.34, it remains an open question whether $\mathrm{NP} \cap \mathrm{co} \cdot \mathrm{NPSV}\text{-sel} \subseteq L_{\Sigma_2^p}$.

4. Hardness for Complexity Classes

4.1 Can P-Selective Sets Be Hard for Complexity Classes?

Let $\leq_r$ be any reducibility and let $\mathcal{C}$ be any complexity class. Recall that a set A is said to be $\mathcal{C}$-$\leq_r$-hard if, for each set $B \in \mathcal{C}$, it holds that $B \leq_r A$. If $A \in \mathcal{C}$ and A is $\mathcal{C}$-$\leq_r$-hard, then we say that A is $\mathcal{C}$-$\leq_r$-complete. When $\leq_r$ is $\leq_m^p$ we will simply write $\mathcal{C}$-hard and $\mathcal{C}$-complete.

In this chapter we ask, for various reductions $\leq_r$ and complexity classes $\mathcal{C}$, whether any P-selective set can be $\mathcal{C}$-$\leq_r$-hard or $\mathcal{C}$-$\leq_r$-complete. Very briefly summarized, the answer we obtain is:

No, unless surprising complexity class collapses occur.

This provides intuitive evidence that the P-selective sets are "simple." For example, Theorem 1.6 shows that if there is an NP-complete P-selective set then P = NP. Since the proof we gave of Theorem 1.6 works equally well if one assumes merely NP-hardness (rather than assuming NP-completeness), the following holds.

Theorem 4.1 P = NP *if and only if there is an NP-hard (i.e., NP-$\leq_m^p$-hard) P-selective set.*

In fact, for the rest of this chapter, we will focus on hardness rather than completeness, since this yields slightly stronger results in the nontrivial directions of theorems such as Theorem 4.1.

The outline of this chapter is as follows. Section 4.2 uses an approach known as the minimum path technique to give evidence that many classes $\mathcal{C}$—UP, Δ_2^p, PSPACE, and EXP—lack $\mathcal{C}$-$\leq_{tt}^p$-hard P-selective sets. Section 4.3 studies whether P-selective sets can be NP-$\leq_{tt}^p$-hard. Section 4.4 gives evidence that even nondeterministically selective sets are unlikely to be hard for standard complexity classes.

4.2 Can P-Selective Sets Be Truth-Table-Hard for UP, Δ_2^p, PSPACE, or EXP?

The minimum path technique is a simple technique that has been extremely successful in showing that P-selective sets are unlikely to be truth-table-hard for various complexity classes. The source of the technique's name can be seen in Lemma 4.3, which captures the insight that focusing on the lexicographically first accepting path may be a route to hardness results. Also central in the success of the technique is the fact that P-selectivity allows groups of strings to be easily ordered in terms of the relationship "is at least as likely to be in the set." Lemma 4.5 (and in some sense Lemma 4.4 also), to be stated soon, captures this property.

Definition 4.2 *Let N be a nondeterministic polynomial-time Turing machine that for some polynomial p has the property that, on each input x, all its computation paths have exactly $p(|x|)$ computation steps. Let our model be such that every step involves a nondeterministic guess bit (though perhaps it is ignored). The length of a computation path will be its number of steps or, equivalently given the above, its number of nondeterministic guess bits. Define $MinimumPath_N = \{\langle x, i \rangle \mid x \in L(N)$ and the lexicographically minimum accepting path of $N(x)$ is of length at least i and has 1 as its ith nondeterministic guess bit\}.*

Regarding Definition 4.2 note that, for strings x such that $x \notin L(N)$, for no i will it be the case that $\langle x, i \rangle \in MinimumPath_N$. The "length at least i" clause is not too crucial here; its function is just to keep things clear as to the case $i > p(|x|)$.

Lemma 4.3 *Let N be a nondeterministic machine as described in Definition 4.2. If there is a P-selective set B such that $MinimumPath_N \leq_{tt}^p B$, then $L(N) \in$ P.*

As mentioned above, to prove Lemma 4.3, we will have to look at one of the most beautiful properties of P-selective sets: They impose polynomial-time computable likelihood orders on all finite sets of strings. By this we mean the following. (As is standard, a k-permutation is simply any one-to-one mapping from $\{1, 2, \ldots, k\}$ onto $\{1, 2, \ldots, k\}$.)

Lemma 4.4 *Let A be any P-selective set. There is a polynomial-time computable function, order, such that*

1. *$(\forall k \geq 1)(\forall x_1, \ldots, x_k)[order(\langle x_1, \ldots, x_k \rangle)$ outputs a permutation of x_1, $\ldots, x_k$, i.e., it outputs $\langle x_{\sigma(1)}, \ldots, x_{\sigma(k)} \rangle$ for some k-permutation σ that may depend on $\langle x_1, \ldots, x_k \rangle]$, and*
2. *$(\forall k \geq 1)(\forall x_1, \ldots, x_k)[(A \cap \{x_1, \ldots, x_k\} = \emptyset) \vee (\exists i : 1 \leq i \leq k)[A \cap \{x_1, \ldots, x_k\} = \{x_{\sigma(1)}, \ldots, x_{\sigma(i)}\}]]$.*

That is, informally, the *order* function orders any set of inputs in such a way that those inputs that are in the P-selective set form a prefix of the order.

Proof of Lemma 4.4 We start by noting a standard fact from tournament theory that is closely related to observations made in Chapter 2. A q-tournament is a directed q-clique, that is, it is a graph having q nodes, no self-loops, and between each pair of distinct nodes i and j either there is a directed edge from i to j or there is a directed edge from j to i—and at most one of these two edges is present. Note that in any q-tournament ($q > 0$) there is at least one node from which every other node can be reached (along a directed path). One way to see this is by a very easy induction. Clearly it holds for 1-tournaments. Suppose, inductively, it holds for all q-tournaments. Consider a $q + 1$-tournament Q. Let x be any fixed node of Q. Let Q' be the q tournament induced by deleting x (and all edges incident on x) from Q. By induction, there is a node y in Q' that can reach each node in Q' via a directed path. Note that if in Q the edge between y and x points toward y, then all nodes in Q can be reached via directed paths from x, and otherwise all nodes in Q (including x) can be reached via directed paths from y. (Another way to see that in any q-tournament there is at least one node from which every other node can be reached along directed paths is that this claim is already clear from Section 2.2.1—since the proof of Lemma 2.6 is essentially a proof of the stronger claim that in any q-tournament there is a node that can reach every other node via directed paths of length at most 2.)

So, we have just noted that in any tournament there is a node that can reach any other node via directed paths. Note further that this is not merely an existence claim; the above proof implicitly gives an algorithm, running in time polynomial in the size of the tournament, for finding such a node. (Regarding the parenthetical remark above that there is a node that can even reach all other nodes via paths of length at most two, one can (given a tournament as a graph) in fact find even such a node in time polynomial in the size of the tournament, since it is a fact from tournament theory that every node of maximum out-degree is such a node. Alternatively, note that even a brute-force search for such a node would run in time polynomial in the size of the tournament.)

Let A be a P-selective set. Let f be a P-selector function for A and, by Theorem 1.4, without loss of generality assume f is a symmetric function (which, recall, means that for any strings x and y it holds that $f(x, y) = f(y, x)$).

We will now define, for the set A, the *order* function. In particular, consider $order(\langle x_1, \ldots, x_k \rangle)$. Without loss of generality assume that the x_i's are distinct, i.e., that $(\forall i \neq j : 1 \leq i \leq k \land 1 \leq j \leq k)[x_i \neq x_j]$. Consider the k-tournament that has one node for each argument of *order*, and such that there is an arrow from the node corresponding to x_i to the node corresponding to x_j if and only if $i \neq j$ and $f(x_i, x_j) = x_j$. Using this tournament, we will define a permutation σ that will dictate the output of $order(\langle x_1, \ldots, x_k \rangle)$.

So, consider the above-defined directed graph, which in particular happens to be a k-tournament. As noted above, for at least one j, it must hold that every node can be reached via directed paths from x_j, and we can in polynomial time find such a j. Choose and fix one such value, j. Set $\sigma(k) = j$. Note that if $x_j \in A$ then certainly each x_i must be in A, by the definition of P-selectivity, the fact that f is a selector for A, the definition of this tournament, and the transitivity of $\leq$ (as applied in a chain to our set's characteristic function on the strings in the paths from x_j to each x_i).

Now, consider the $(k-1)$-tournament formed by deleting from the k-tournament x_j and all edges touching x_j. As noted above, for some ℓ (if $k - 1 \geq 1$) it must hold that every node in this $(k-1)$-tournament can be reached via directed paths from x_ℓ, and we can easily find and fix one such value, call it ℓ. Set $\sigma(k-1) = \ell$. Note that if $x_\ell \in A$ then certainly all x_i except perhaps x_j must be in A.

Continue this process until the permutation σ is completely defined. Note that this algorithm indeed outputs a permutation having the desired property. Furthermore, the above is a polynomial-time algorithm, since as noted above the "finding a node from which all other nodes can be reached" step is easily a polynomial-time step, and f is a polynomial-time function.   ❑   Lemma 4.4

Note that neither the statement nor the proof of Lemma 4.4 addresses the issue of whether there is some selector function, f, having the property that after the reordering we know that the ordering not only makes the elements of the set a left-segment of the output reordering, but also respects the selector function in the sense that for each $i < k$ we have $f(x_{\sigma(i)}, x_{\sigma(i+1)}) = x_{\sigma(i)}$. In fact, we can use a different $order$ function to achieve that type of behavior relative to any given symmetric selector function.

Lemma 4.5 *Let A be any P-selective set, and let f be any symmetric P-selector function for A. Then there is a polynomial-time computable function, order, such that*

1. $(\forall k \geq 1)(\forall x_1, \ldots, x_k)[order(\langle x_1, \ldots, x_k \rangle)$ *outputs a permutation of x_1, $\ldots, x_k$, i.e., it outputs $\langle x_{\sigma(1)}, \ldots, x_{\sigma(k)} \rangle$ for some k-permutation σ that may depend on $\langle x_1, \ldots, x_k \rangle$], and*
2. $(\forall k \geq 1)(\forall x_1, \ldots, x_k)[(A \cap \{x_1, \ldots, x_k\} = \emptyset) \vee (\exists i : 1 \leq i \leq k)[A \cap \{x_1, \ldots, x_k\} = \{x_{\sigma(1)}, \ldots, x_{\sigma(i)}\}]]$.
3. $(\forall k \geq 1)(\forall x_1, \ldots, x_k)(\forall i : 1 \leq i < k)[f(x_{\sigma(i)}, x_{\sigma(i+1)}) = x_{\sigma(i)}]$.

Proof   We merely quickly sketch the proof. The idea is, basically, to mimic bubblesort. Take x_1. Make it the initial, tentative left-most element of the output. Take x_2. Put it to the left of x_1 if $f(x_1, x_2) = x_2$ and to the right of x_1 otherwise. Now we have ordered two elements. Take the next element.

Start it on the far left and percolate it to the right until the first time it beats the element to its right, at which point stop and put it just to the left of the element that it beat. Note that at this point it beats (with respect to the selector function) the element, if any, to its right, and it loses (with respect to the selector function) to the element, if any, to its left. Continue until all the elements have been thusly inserted. The element leftmost at the end of this process becomes our $x_{\sigma(1)}$, the next to leftmost becomes $x_{\sigma(2)}$, and so on.   □

Since whether one wants our ordering lemmas in a "left-prefix" version or a "right-suffix" version is a matter of taste, we here state the "right-suffix" versions of Lemmas 4.4 and 4.5.

Lemma 4.6 *Let A be any P-selective set. There is a polynomial-time computable function, order, such that*

1. $(\forall k \geq 1)(\forall x_1, \ldots, x_k)[order(\langle x_1, \ldots, x_k \rangle)$ *outputs a permutation of x_1, $\ldots, x_k$, i.e., it outputs $\langle x_{\sigma(1)}, \ldots, x_{\sigma(k)} \rangle$ for some k-permutation σ that may depend on $\langle x_1, \ldots, x_k \rangle]$, and*

2. $(\forall k \geq 1)(\forall x_1, \ldots, x_k)[(A \cap \{x_1, \ldots, x_k\} = \emptyset) \vee (\exists i : 1 \leq i \leq k)[A \cap \{x_1, \ldots, x_k\} = \{x_{\sigma(i)}, \ldots, x_{\sigma(k)}\}]]$.

Lemma 4.7 *Let A be any P-selective set, and let f be any symmetric P-selector function for A. Then there is a polynomial-time computable function, order, such that*

1. $(\forall k \geq 1)(\forall x_1, \ldots, x_k)[order(\langle x_1, \ldots, x_k \rangle)$ *outputs a permutation of x_1, $\ldots, x_k$, i.e., it outputs $\langle x_{\sigma(1)}, \ldots, x_{\sigma(k)} \rangle$ for some k-permutation σ that may depend on $\langle x_1, \ldots, x_k \rangle]$, and*

2. $(\forall k \geq 1)(\forall x_1, \ldots, x_k)[(A \cap \{x_1, \ldots, x_k\} = \emptyset) \vee (\exists i : 1 \leq i \leq k)[A \cap \{x_1, \ldots, x_k\} = \{x_{\sigma(i)}, \ldots, x_{\sigma(k)}\}]]$.

3. $(\forall k \geq 1)(\forall x_1, \ldots, x_k)(\forall i : 1 \leq i < k)[f(x_{\sigma(i)}, x_{\sigma(i+1)}) = x_{\sigma(i+1)}]$.

With our ordering lemma (Lemma 4.4) in hand, we can prove our main "minimum path" lemma.

Proof of Lemma 4.3   Let N be a nondeterministic machine as described in Definition 4.2. Let B be a P-selective set such that $MinimumPath_N \leq_{tt}^p B$.

We now give a polynomial-time algorithm for $L(N)$. Let p be as in Definition 4.2, i.e., p is the polynomial giving the lengths of paths of machine N. For each j, $1 \leq j \leq p(|x|)$, consider the question "$\langle x, j \rangle \in MinimumPath_N$?", which must truth-table reduce to B. Compute the list of truth-table questions to B that this reduction asks on input $\langle x, 1 \rangle$, and compute the list of truth-table questions to B that this reduction asks on input $\langle x, 2 \rangle$, and so on, through the list of truth-table questions to B that this reduction asks on input $\langle x, p(|x|) \rangle$. Compute the union of all queries from these lists. Since we are unioning a polynomial number of polynomial-sized lists, the union is itself of polynomial size. Furthermore, given the answers to all the queries on this long list, we can (using the truth-table reduction) easily compute for each j whether $\langle x, j \rangle \in MinimumPath_N$.

Suppose our giant list is $y_1, \ldots, y_k$, where as just noted k varies with the input x but is polynomially bounded in $|x|$. By Lemma 4.4 obtain a permutation—$y_{\sigma(1)}, \ldots, y_{\sigma(k)}$—of the y_i's such that the permutation satisfies the properties of Lemma 4.4. Though k queries usually have 2^k possible answers, thanks to Lemma 4.4 there are only $k + 1$ possible answers to this set of k queries. It is possible that all the y_i's are answered no. It is possible that $y_{\sigma(1)}$ is answered yes and all the other queries are answered no. It is possible that $y_{\sigma(1)}$ and $y_{\sigma(2)}$ are answered yes and all the other queries are answered no. And so on, up to the possibility that all queries are answered yes. Each of these $k + 1$ possibilities gives an answer to all of the questions $\langle x, j \rangle \in MinimumPath_N$ (for $1 \leq j \leq p(|x|)$). One of the $k + 1$ possibilities actually gives the *correct* answers.

Unfortunately, we do not know which of the $k + 1$ possibilities gives the correct answers. However, we can proceed as follows. For each of the $k + 1$ possibilities in turn, suppose that that possibility is the one giving the correct answers to the queries to B. Then these answers in turn give us answers to the questions $\langle x, j \rangle \in MinimumPath_N$, for $1 \leq j \leq p(|x|)$. Consider the computation path specified by these $p(|x|)$ bits. If it is an accepting path, then certainly $x \in L(N)$, since $N(x)$ has an accepting path. If it is not an accepting path, go on to the next of the $k+1$ possibilities, and check whether it yields an accepting path, and so on.

If any of the $k + 1$ possibilities yields an accepting path, then we will accept, and this is correct, since we have found an accepting path, and thus have found a proof that $x \in L(N)$. If each of the $k + 1$ possibilities yields rejecting paths, then we will reject. This is also correct, as can be seen as follows. If $x \in L(N)$, then there is some lexicographically minimum accepting path, and so the true answers to the k queries on our list would have given its bits. However, we tried $k + 1$ options of which one was guaranteed to be the correct answer sequence and none of the tried options gave the bits of an accepting path.

This completes our polynomial-time algorithm for $L(N)$.

❑   Lemma 4.3

With Lemma 4.3 in hand, the main results about $\leq_{tt}^p$-hardness now follow easily.

Theorem 4.8

1. $P = UP$ if and only if there is a UP-$\leq_{tt}^p$-hard P-selective set.
2. $P = NP$ if and only if there is a Δ_2^p-$\leq_{tt}^p$-hard P-selective set.
3. $P = PSPACE$ if and only if there is a $PSPACE$-$\leq_{tt}^p$-hard P-selective set.
4. $EXP \not\subseteq R_{tt}^p(\text{P-sel})$.

Proof The three "only if" directions are all trivial (e.g., if $P = UP$ then the finite, P-selective set $\{1\}$ is even $\leq_m^p$-complete for UP).

Let us turn to the "if" direction of part 1. Let B be a P-selective set that is UP-$\leq_{tt}^p$-hard. Let A be an arbitrary set in UP. We say a Turing machine N is unambiguous (see the entry for UP in Appendix A.2) if and only if, for all x, it holds that N on input x has at most one accepting computation path. So, there is an unambiguous polynomial-time Turing machine, N', accepting A. Note that there will exist a machine N that is also an unambiguous polynomial-time Turing machine accepting A, that has the properties required by Definition 4.2. Note that for this machine N, it holds that $MinimumPath_N \in$ UP. Since B is UP-$\leq_{tt}^p$-hard and $MinimumPath_N \in$ UP, clearly $MinimumPath_N \leq_{tt}^p B$. So by Lemma 4.3 we have $L(N) \in$ P, i.e., $A \in$ P.

The "if" direction of part 2 is quite analogous, except that for an arbitrary NP machine, all one can claim—at least as far as is currently known—is that its "$MinimumPath$" set is in Δ_2^p.

Regarding the "if" direction of part 3, let B be a P-selective set that is PSPACE-$\leq_{tt}^p$-hard. Since R_{tt}^p(P-sel) $\subseteq R_T^p$(P-sel) $=$ P/poly (see Proposition 2.25), by Theorem 2.27 we have PSPACE $= \Sigma_2^p$. Since $\Delta_2^p \subseteq$ PSPACE, by part 2 of the current theorem, P $=$ NP (and thus P $= \Sigma_2^p$). So P $= \Sigma_2^p =$ PSPACE.

Finally, we turn to part 4. Assume that EXP $\subseteq R_{tt}^p$(P-sel). From Proposition 2.25 and Theorem 2.28, EXP $= \Sigma_2^p$. Since PSPACE $\subseteq$ EXP, part 3 of the current theorem implies P $=$ PSPACE. So P $=$ PSPACE $\subseteq$ EXP $= \Sigma_2^p \subseteq$ PSPACE $=$ P, and thus P $=$ EXP. However, P $\neq$ EXP (this is one consequence of the deterministic time hierarchy theorem), and so we have contradicted our assumption that EXP $\subseteq R_{tt}^p$(P-sel). $\square$

Note, however, that part 2 of Theorem 4.8 leaves open whether it holds that P $=$ NP if and only if there is a NP-$\leq_{tt}^p$-hard P-selective set. Section 4.3 will be devoted to seeking consequences of NP having $\leq_{tt}^p$-hard P-selective sets.

4.3 Can P-Selective Sets Be Truth-Table-Hard or Turing-Hard for NP?

Can any P-selective set be NP-$\leq_{tt}^p$-hard or NP-$\leq_T^p$-hard? Let us first briefly consider the $\leq_T^p$ case. By Proposition 2.25, this is equivalent to the issue of whether any P/poly set is NP-$\leq_T^p$-hard. Thus, Theorem 2.29 (in light of Proposition 2.25) gives the following result.

Theorem 4.9 *If* NP $\subseteq R_T^p$(P-sel) *then* PH $=$ ZPP^{NP}.

In fact, one can claim a bit more, in light of Theorem 2.29 and the results we know about the nonuniform containments that apply to nondeterministically selective sets. (At the end of the Bibliographic Notes for this chapter we mention some very recent work that strengthens, even beyond the collapses stated in Theorem 4.9 and Corollary 4.10, the consequences that follow from the hypotheses of those theorems.)

Corollary 4.10

1. *If* $NP \subseteq R_T^p(NPSV_t\text{-sel})$ *then* $PH = ZPP^{NP}$.
2. *If some* $NP\text{-}\leq_T^p\text{-complete set is }NPSV\text{-sel, then }PH = ZPP^{NP}$.

Proof It is not hard to see that $P^{(NP \cap coNP)/poly} = (NP \cap coNP)/poly$. So the claims hold in light of Theorem 2.29, due to Theorems 2.18 and 2.19. ☐

We now turn to whether any P-selective set can be NP-$\leq_{tt}^p$-hard. It is an open question whether $NP \subseteq R_{tt}^p(P\text{-sel})$ implies $P = NP$. In fact, this is probably the most intensely studied open question in P-selectivity theory. A large number of weaker results are currently known to hold. In particular, Theorem 4.11 shows that weaker conclusions follow from the desired hypothesis, and Theorem 4.12 shows that the desired conclusion follows from a stronger hypothesis. Below, FP_{tt}^{NP} denotes the class of functions computable via polynomial-time truth-table access to NP; $FP^{NP[\mathcal{O}(\log n)]}$ denotes the class of functions computable in polynomial time using $\mathcal{O}(\log n)$ Turing queries to NP. That is, these are function analogs of $R_{tt}^p(NP)$ and $R_{\mathcal{O}(\log n)\text{-}T}^p(NP)$. Though $R_{tt}^p(NP) = R_{\mathcal{O}(\log n)\text{-}T}^p(NP)$, it remains an open issue whether the function analogs of the classes are equal. Indeed, Lemma 4.13 will suggest that equality is unlikely.

Theorem 4.11 *If* $NP \subseteq R_{tt}^p(P\text{-sel})$ *then each of the following holds:*

1. $PH = ZPP^{NP}$.
2. $P = FewP$.
3. $R = NP$.
4. $FP_{tt}^{NP} \subseteq FP^{NP[\mathcal{O}(\log n)]}$.
5. $SAT \in \bigcap_{k>0} DTIME[2^{n/\log^k n}]$.

Theorem 4.12 *If* $NP \subseteq R_{btt}^p(P\text{-sel})$ *then* $P = NP$.

The proofs of the parts of Theorem 4.11 differ sharply from each other, and so we prove them separately. Part 1 follows immediately from Theorem 4.9. We will leave part 3 unproven; see the Bibliographic Notes for references to its proof. We now prove part 2 of Theorem 4.11. The minimum path technique was discussed in detail in Section 4.2. The proof of this part of the theorem relies on a variation of the minimum path technique. In particular, it also employs "the parallel census technique"—the fact that by trying in parallel all possible numbers of accepting paths, via a truth-table call containing polynomially many queries to NP one can obtain a list of all accepting paths of a FewP machine on a given input.

Proof of Part 2 of Theorem 4.11 Assume that $NP \subseteq R_{tt}^p(P\text{-sel})$. In particular, let B be a P-selective set that is NP-$\leq_{tt}^p$-hard. Let A be an arbitrary set in FewP. So there will be a nondeterministic polynomial-time Turing machine, N', accepting A such that N' has a polynomially bounded number of accepting paths. It is not hard to see that there will exist a machine N that is also a polynomial-time Turing machine accepting A, that on each input has the same number of accepting paths as N', and that has the properties required by Definition 4.2. Let $MinimumPath'_N = \{\langle 0^k, x, i, j \rangle \mid x \in L(N)$ and $i \leq k$ and there exist k distinct accepting paths of $N(x)$, $p_1 <_{lex} p_2 <_{lex} \cdots <_{lex} p_k$, such that p_i has at least j nondeterministic guess bits and its jth nondeterministic guess bit is $1\}$. Note that for our machine N it holds that $MinimumPath'_N \in NP$. Since B is NP-$\leq^p_{tt}$-hard and $MinimumPath'_N \in NP$, clearly $MinimumPath'_N \leq^p_{tt} B$.

We now give a polynomial-time algorithm for $A = L(N)$. Let p be the polynomial (related to N) of Definition 4.2, i.e., $p(|x|)$ is the length (equivalently, the number of nondeterministic guess bits) of the computation paths of $N(x)$. Let q be a polynomial such that, for each input x, it holds that $q(|x|)$ is greater than or equal to the number of accepting paths of $N(x)$. The polynomial-time algorithm for $L(N)$ works as follows.

For each k, $1 \leq k \leq q(|x|)$, do the following procedure (which is a more sophisticated version of the procedure used in the proof of Lemma 4.3). In each of the $q(|x|)$ invocations of the following procedure, some polynomial number of truth-table queries intended for B will be generated.

Consider the questions "$\langle 0^k, x, i, j \rangle \in MinimumPath'_N$?" for each $1 \leq i \leq k$ and each $1 \leq j \leq p(|x|)$. Each of these questions truth-table reduces to B. Compute the list of all truth-table questions to B amongst all these $kp(|x|)$ queries, i.e., union together all the query sets. Since we are combining a polynomial number of polynomial-sized lists, the combined list is itself of polynomial size. Furthermore, given answers to all queries on this giant list, we can (using the truth-table reduction) easily compute for each i and j whether $\langle 0^k, x, i, j \rangle \in MinimumPath'_N$. Suppose our giant list is $y_1, \ldots, y_z$, where as just noted z varies with the input x but is polynomially bounded in $|x|$. By Lemma 4.4 obtain a permutation—$y_{\sigma(1)}, \ldots, y_{\sigma(z)}$—of the y_i's such that the permuted order satisfies the properties of Lemma 4.4. Though z queries usually have 2^z possible answers, by exactly the same argument as in Lemma 4.3 we can come up with a set of $z + 1$ possible answers strings to this set of queries.[9] Each of these $z + 1$ possibilities gives an answer to each of the $kp(|x|)$ questions, "$\langle 0^k, x, i, j \rangle \in MinimumPath'_N$?" One of the $z+1$ possibilities actually gives the correct answers. Unfortunately, we do not know which of the $z + 1$ possibilities gives the correct answers. Nonetheless, we can proceed as follows. For each of the $z + 1$ possibilities in turn, suppose that possibility is the one giving the correct answers to the queries to B.

---

[9] That is, by Lemma 4.4 there are only $z+1$ possible answers to this set of queries. Namely, it is possible that all the y_i's are answered no. It is possible that $y_{\sigma(1)}$ is answered yes and all the other queries are answered no. It is possible that $y_{\sigma(1)}$ and $y_{\sigma(2)}$ are answered yes and all the other queries are answered no. And so on, up to the possibility that all the queries are answered yes.

Then these answers in turn give us answers to the questions "$\langle 0^k, x, i, j \rangle \in$ $MinimumPath'_N$?", for $1 \leq i \leq k$ and $1 \leq j \leq p(|x|)$. These answers give k computation paths. If any one of them is an accepting path then accept. Otherwise, move on to the next value of k.

If for no k, $1 \leq k \leq q(|x|)$, does the above procedure yield an accepting path, then reject.

We must now argue that the above algorithm, which clearly runs in polynomial time, accepts A. The above algorithm accepts x only if the algorithm has found an actual accepting path. Thus, if the algorithm accepts, then certainly $x \in A$. Regarding the other direction, suppose $x \in A$. So $N(x)$ accepts, and has at most $q(|x|)$ accepting paths. Let k' be the number of accepting paths of $N(x)$. Clearly, $k' \leq q(|x|)$. Note that the above procedure, when run on the value $k = k'$, will have the property that the answers to the $MinimumPath'_N$ queries give exactly the bits of each of the k' accepting paths of $N(x)$. Thus the $k = k'$ iteration of the above loop will find an accepting path of $N(x)$ and thus will accept (unless some earlier value of k already caused us to accept, which is also fine). Thus, the algorithm accepts x if and only if $x \in A$.   ❑   Part 2 of Theorem 4.11

The proof of part 4 of Theorem 4.11 draws upon the ordering lemma (Lemma 4.4) and a binary search procedure.

Proof of Part 4 of Theorem 4.11   Let f be an arbitrary function computable in FP_{tt}^{NP}. We seek to prove $f \in FP^{NP[\mathcal{O}(\log n)]}$, under the hypothesis of the theorem. In particular, assume that $NP \subseteq R_{tt}^p(\text{P-sel})$. Let B be a P-selective set that is NP-$\leq_{tt}^p$-hard. Let A be an NP set such that $f \in FP_{tt}^A$. Since B is NP-$\leq_{tt}^p$-hard, certainly $A \leq_{tt}^p B$.

We now give a $FP^{NP[\mathcal{O}(\log n)]}$ algorithm for f. Let x be the input. Fix some particular truth-table reduction showing that $f \in FP_{tt}^A$. Compute the list of questions, $x_1, \ldots, x_z$, that the truth-table reduction generates to be asked to A on input x. Each of the z queries itself truth-table reduces to B. Union together the z sets of queries to B created by reducing from each x_i to B. Call this list's members $y_1, \ldots, y_\ell$. Invoking Lemma 4.4, obtain a permutation σ obeying the conditions of that lemma. Consider $y_{\sigma(1)}, \ldots, y_{\sigma(\ell)}$. Though ℓ membership queries usually have 2^ℓ plausible answers, thanks to Lemma 4.4 again there are only $\ell+1$ plausible answers to this set of queries. (It is possible that all the y_i's are answered no. It is possible that $y_{\sigma(1)}$ is answered yes and all the other queries are answered no. It is possible that $y_{\sigma(1)}$ and $y_{\sigma(2)}$ are answered yes and all the other queries are answered no. And so on, up to the possibility that all queries are answered yes.) Each of these $\ell+1$ possibilities gives (via the truth-table reduction) an answer to each of the ℓ questions to B and so to each of the z questions to A. At least one of the $\ell+1$ possibilities actually gives (via the truth-table reduction) the *correct* answers to the z questions to A, namely the one that corresponds to the correct answers to the queries to B. As always, we do not know what the correct answers to the queries to B are. To come up with the correct answers, we can, however, proceed as follows. For each of the $\ell + 1$ possibilities in turn, consider the set of answers to the z questions to A given by that particular possibility. If all $\ell + 1$ possibilities yield exactly the same answers to the z queries, then we know the correct answers to the queries to A, and are done without having to make any queries to an NP oracle.

On the other hand, suppose that the $\ell + 1$ possibilities don't all give the same answers to the z questions to A. Then, for each of the $\ell + 1$ possibilities, see what answer the possibility gives to the A queries. Consider the set of all such sets of answers to the A queries. We have at most $\ell + 1$ (but possibly fewer) set of answers to the A queries. View each set of answers as a bitstring (the first bit being the answer to the first question to A, and so on), and sort the bitstrings in lexicographic order. Now, by binary search, we will determine which of these bitstrings in fact contains the true answers to the queries to A. We do so as follows.

Consider the NP set

$$C = \{\langle 1, k, \langle x_1, \ldots, x_n \rangle\rangle \mid \|A \cap \{x_1, \ldots, x_n\}\| \geq k\}$$
$$\cup \; \{\langle 0, \langle s_1, \ldots, s_m \rangle, \langle x_1, \ldots, x_n \rangle\rangle \mid (\exists i)[yes(s_i, \langle x_1, \ldots, x_n \rangle)]\},$$

where $yes(s_i, \langle x_1, \ldots, x_n \rangle)$ holds exactly if $(\forall j : 1 \leq j \leq |s_i|)[$ (the jth bit of s_i is 1) $\implies x_j \in A]$.

Our binary search works as follows. Via $\mathcal{O}(\log |x|)$ sequential queries to C of the form $\langle 1, k, \langle x_1, \ldots, x_z \rangle\rangle$, with k varying in a binary search fashion, determine $\|A \cap \{x_1, \ldots, x_z\}\|$. Let that value be denoted e. Now, we know exactly how many of the strings $x_1, \ldots, x_z$ belong to A. Among the at most $\ell + 1$ possibilities regarding the answers to A, discard all those that do not say that exactly e elements belong to A. Let $s_1, \ldots, s_w$ denote those bitstrings claiming that exactly e elements of $x_1, \ldots, x_z$ belong to A. With $\mathcal{O}(\log |x|)$ additional queries to C, we can determine exactly which s_i is the correct set of answers to A. For example, we first ask of C the query $\langle 0, \langle s_1, \ldots s_{\lfloor w/2 \rfloor} \rangle, \langle x_1, \ldots, x_z \rangle\rangle$. If the answer is yes (respectively, no), then we ask about $\langle 0, \langle s_1, \ldots s_{\lfloor w/4 \rfloor} \rangle, \langle x_1, \ldots, x_z \rangle\rangle$ (respectively, $\langle 0, \langle s_{1+\lfloor w/2 \rfloor}, \ldots s_{\lfloor 3w/4 \rfloor} \rangle, \langle x_1, \ldots, x_z \rangle\rangle$). We continue this process until we pinpoint exactly which s_i gives the correct answers to the queries to A. Knowing the correct answers to all queries to A on input x, we can now easily (by simulating what the FP_{tt}^A machine does on input x given those answers to $x_1, \ldots, x_z$) compute $f(x)$. Since we have done so using at most $\mathcal{O}(\log |x|)$ queries to the NP set C, this shows that $f \in \text{FP}^{\text{NP}[\mathcal{O}(\log n)]}$.   ☐   Part 4 of Theorem 4.11

Part 5 of Theorem 4.11 follows immediately from part 4 of Theorem 4.11 in light of the following lemma, which we state without proof.

Lemma 4.13 *If* $\text{FP}_{tt}^{\text{NP}} \subseteq \text{FP}^{\text{NP}[\mathcal{O}(\log n)]}$ *then*

$$\text{SAT} \in \bigcap_{k>0} \text{DTIME}[2^{n/\log^k n}].$$

We now turn to the proof of Theorem 4.12.

Proof of Theorem 4.12   Assume NP $\subseteq$ R$^p_{btt}$(P-sel). Let $k \geq 0$ and $A \in$ P-sel be such that SAT $\leq^p_{k\text{-}tt} A$, and without loss of generality let $k \geq 5$ (by asking dummy questions, if needed). We will use SAT $\leq^p_{k\text{-}tt} A$ to build a deterministic polynomial-time algorithm for SAT. Our general approach will be based on the proof of Theorem 1.6. Recall that, there, to give a deterministic polynomial-time algorithm for SAT, we repeatedly self-reduced a formula—in particular, we took an unassigned variable of the formula and obtained two formulas by assigning to the variable the values true and false—and then pruned away one of the two resulting formulas.

In the present setting, we will have to be a bit less aggressive in our pruning. At each state, we will start with a set of at most $2(2^k - 1)$ prefixes of assignments to our input formula, and will prune this set until we have at most $2^k - 1$ prefixes of assignments to our input formula. We will do so in such a way that if at least one of the at most $2(2^k - 1)$ prefixes we started the stage with is a prefix of a satisfying assignment, then at least one of the at most $2^k - 1$ prefixes we end the stage with will be a prefix of a satisfying assignment. At the end of each stage, we will take each of the at most $2^k - 1$ prefixes we have and turn each into two prefixes by assigning the next unassigned variable both true and false.

At stage 0, we will start with one assignment prefix for the input boolean formula $F(x_1, \ldots, x_m)$, namely, the empty assignment. At the start of the (special) final stage, we will have assigned all the variables, and thus will have a set of at most $2(2^k - 1)$ complete variable assignments such that $F \in$ SAT if and only if at least one of these at most $2(2^k - 1)$ assignments satisfies F. However, this means that $F \in$ SAT is now easy to check in polynomial-time, via simply trying each of these assignments.

We have now sketched the entire algorithm, except how the pruning is done. Let us turn to that. So, we have a collection of at most $2(2^k - 1)$ prefixes of assignments of F, and we wish to prune down to at most $2^k - 1$ prefixes of assignments in such a way as to not eliminate all prefixes of satisfying assignments, if there are any such. If we have at most $2^k - 1$ prefixes of assignments coming into the stage, no pruning is needed, so we proceed to the next stage. Otherwise, pruning is needed. We will show how to, in polynomial time, appropriately prune one prefix from any set of 2^k prefixes of assignments to F. By using this repeatedly, we can in polynomial time prune from at most $2(2^k - 1)$ prefixes to at most $2^k - 1$ prefixes.

Let our 2^k prefixes of assignments to F be $\alpha_1, \alpha_2, \ldots, \alpha_{2^k}$. Let $F[\alpha_i]$ denote the formula obtained when making the assignment α_i in the formula F. Define the following formulas, $1 \leq j \leq k$:

$$G_j = \bigvee_{\substack{i \,:\, 1 \leq i \leq 2^k \text{ and the } j\text{th} \\ \text{bit of } i - 1 \text{ (in binary) is 1}}} F[\alpha_i].$$

Note that for each j, $1 \leq j \leq k$,

$$G_j \in \text{SAT} \iff (\exists i : 1 \leq i \leq 2^k)[(\text{the } j\text{th bit of } i-1 \text{ is } 1) \wedge (F[\alpha_i] \in \text{SAT})].$$

Crucially, there are only a very small number of possibilities regarding which collections of G_j's may be satisfiable. To see this, note that since SAT $\leq^p_{k\text{-}tt} A$ each question "$G_j \in \text{SAT}$?" polynomial-time k-truth-table reduces to A. So, over all the G_j, we get a total of at most k^2 queries to A. Each query gets either a yes or a no answer, so we might worry that there are about 2^{k^2} ways these queries could be answered. Fortunately, A is P-selective, so we can do *much* better. In particular, we apply Lemma 4.4 to these k^2 query strings. This puts them into an order such that either all are in $\overline{A}$, or the first is in A and the rest are in $\overline{A}$, or the first two are in A and the rest are in $\overline{A}$, or ..., or all but the last one are in A and the last one is in $\overline{A}$, or all are in A. So, we obtain a group of at most $k^2 + 1$ ways of answering all k queries ("$G_j \in \text{SAT}$?", $1 \leq j \leq k$) such that one of these $k^2 + 1$ ways is correct.

Since $k \geq 5$, we have $k^2 + 1 < 2^k$, so at least one of the 2^k potential satisfiability settings of the G_j's is excluded. Fix some excluded possibility, $b = b_1 b_2 \cdots b_k$, where for $1 \leq j \leq k$ we let $b_j = 1$ if the fixed excluded possibility has $G_j \in \text{SAT}$ and we let $b_j = 0$ if the fixed excluded possibility has $G_j \notin \text{SAT}$.

Now, we can finally prune one assignment prefix, as was our goal. In particular, we may safely prune α_{b+1}. Why? The only dangerous case would be if

$$F[\alpha_{b+1}] \in \text{SAT} \wedge (\forall i : 1 \leq i \leq 2^k)[i \neq b+1 \implies F[\alpha_i] \notin \text{SAT}],$$

since in this case (and only in this case) we would be pruning the only prefix leading to a satisfying assignment. However, if $F[\alpha_{b+1}] \in \text{SAT} \wedge (\forall i : 1 \leq i \leq 2^k)[i \neq b+1 \implies F[\alpha_i] \notin \text{SAT}]$, then by the definition of the G_j's we have, for $1 \leq j \leq k$,

$$G_j \in \text{SAT} \iff \text{the } j\text{th bit of } b \text{ is } 1.$$

However, this says that the one excluded possibility (that named by b) in fact is exactly the case that holds, which is a contradiction. Thus, by pruning α_{b+1} we certainly will never prune the only prefix leading to a satisfying assignment. This concludes the construction of a way of pruning one possibility, and concludes our proof.   ❑   Theorem 4.12

4.4 Can Nondeterministically Selective Sets Be NP-Hard or coNP-Hard?

In this section, we will see that nondeterministically selective sets are unlikely to be NP-$\leq^p_m$-hard or coNP-$\leq^p_m$-hard. The main result is the following.

Theorem 4.14 *The results stated in the following table hold.*

$\mathcal{F}$	NP $\subseteq \mathcal{F}$-sel		coNP $\subseteq \mathcal{F}$-sel	
NPSV_t	*holds iff* NP = coNP	(a)	*holds iff* NP = coNP	(b)
NPSV	*holds if* NP = coNP	(c)	*holds iff* NP = coNP	(d)
	holds only if PH = ZPP^{NP}	(e)		
NPMV_t	*holds iff* NP = coNP	(f)	*holds iff* NP = coNP	(g)
NPMV	*holds (without any assumption)*	(h)	*holds iff* NP = coNP	(i)

We will establish each of the nine claims contained in the theorem. In fact, for some parts we will prove results that are, in the nontrivial direction, stronger than the claims in the theorem. For example, part f of Theorem 4.14 says that unless NP = coNP no NPMV_t-selective set is NP-$\leq_m^p$-hard. In fact, we will see that the same claim holds even for NP-$\leq_\gamma$-hardness and NP-$\leq_{1\text{-}tt}^p$-hardness (see Corollary 4.22).

Theorem 4.15 asserts that part h of Theorem 4.14 holds.

Theorem 4.15 NP $\subseteq$ NPMV-sel.

Proof Let L be any NP set. Consider the NPMV function $f_L(x, y)$ defined by

$$\text{set-}f_L(x, y) = \{x, y\} \cap L.$$

f_L is clearly an NPMV-selector for L. ❑

Note that it follows easily from Theorem 4.15 that all the "if" directions of Theorem 4.14 hold.

Lemma 4.16 *Part c, and the "if" directions of Parts a, b, d, f, g, and i of Theorem 4.14 all hold.*

Proof Assume NP = coNP. So by Theorem 4.15 and part 2 of Proposition 1.32 we have coNP = NP $\subseteq$ NPMV-sel = NPSV_t-sel = NPSV-sel = NPMV_t-sel. ❑

Part e of Theorem 4.14 follows immediately from Theorem 2.19 and Theorem 2.29.

Recall from Section 1.1.2 the definition of Turing self-reducibility. We now turn to the "only if" directions of Parts b, d, g, and i of Theorem 4.14. They will all follow from Theorem 4.17, which is of interest in its own right.

Theorem 4.17 *If A is Turing self-reducible and NPMV-selective then $A \in$ NP.*

We defer the proof of Theorem 4.17 to Section 5.4, which focuses on self-reducibility. It will be proven there as Theorem 5.21 after we first establish there a natural, analogous result for P-selectivity.

Corollary 4.18 *The "only if" directions of parts b, d, g, and i of Theorem 4.14 hold.*

Proof By Theorem 4.17 and the fact that SAT (and thus $\overline{\text{SAT}}$) is Turing self-reducible, we have

$$\text{coNP} \subseteq \text{NPMV-sel} \implies \text{NP} = \text{coNP}.$$

This is part i. Parts b, d, and g follow immediately from part i, due to Proposition 1.26. ❏

Note that Theorem 4.17 can be usefully applied to any class containing complete sets that are Turing self-reducible. For example, we can state the following.

Theorem 4.19

1. PSPACE $\subseteq$ NPMV-sel *if and only if* PSPACE = NP.
2. $\Sigma_2^p \subseteq$ NPMV-sel *if and only if* PH = NP.
3. PP $\subseteq$ NPMV-sel *if and only if* PP = NP.

We will now prove the "only if" directions of parts a and f of Theorem 4.14. We will establish them via the following result, which is of interest in its own right. We defer the proof of Theorem 4.20 to Section 5.4, where (as Theorem 5.20) we will see that it is a surprisingly general nondeterministic analog to the standard deterministic results on self-reducibility and P-selective sets.

Theorem 4.20 *If A is Turing self-reducible and $A \in \text{R}_\gamma(\text{R}_{1\text{-}tt}^p(\text{NPMV}_t\text{-sel}))$ then $A \in \text{NP} \cap \text{coNP}$.*

Corollary 4.21 *The "only if" directions of parts a and f of Theorem 4.14 hold.*

Proof Part f is immediate from Theorem 4.20. Part a is immediate from part f in light of Proposition 1.26. ❏

In fact, Theorem 4.20 yields a bit more than the "only if" direction of part f of Theorem 4.14.

Corollary 4.22 *If* NP $\subseteq \text{R}_\gamma(\text{R}_{1\text{-}tt}^p(\text{NPMV}_t\text{-sel}))$ *then* PH = NP. *In particular, no* NPMV_t*-selective set can be* NP-$\leq_{1\text{-}tt}^p$*-hard or* NP-$\leq_\gamma$*-hard unless* PH = NP.

4.5 Bibliographic Notes

The minimum path technique is due to Toda's breakthrough 1991 paper [Tod91]. The ordering insight dates back to Ko's ordering work (see Theorem 1.15), and in the form presented here—Lemma 4.4 and its variants—is most famously due to Toda [Tod91]. The result is sometimes referred to as the Toda Ordering Lemma, or as Toda's Lemma. For a reference regarding the simple fact, noted in passing in the proof of Lemma 4.4, that every maximum out-degree node in a tournament reaches all other nodes via directed paths of length at most two see, for example, West [Wes96]. Definition 4.2, Lemma 4.3, and Theorem 4.8 are due to Toda [Tod91]. Part 2 of Theorem 4.8 was obtained, independently of Toda, by Beigel [Bei88]. The deterministic time hierarchy theorem referred to in the proof of part 4 of Theorem 4.8 is a classic result due to Hartmanis and Stearns [HS65].

Part 4 of Theorem 4.8 has been extended by Burtschick and Lindner [BL97], who prove $(\forall k)[\text{EXP} \not\subseteq \text{E}^{\text{P-sel}[n^k]}]$, that is, that E, given n^k queries to any P-selective set, fails to contain EXP.

Theorem 4.9 is a weakened restatement of part of Theorem 2.29 in light of Proposition 2.25.

Can the conclusion of Theorem 4.9 be strengthened? The answer is yes, but there seem to be limits. In particular, very recently Cai [Cai01] has strengthened the conclusion of Theorem 4.9 to $\text{PH} = \text{S}_2$, and Cai et al. [CCHO01] have strengthened Corollary 4.10's conclusion to $\text{PH} = \text{S}_2^{\text{NP}\cap\text{coNP}}$. There are two relativized results that suggest that Theorem 4.9 cannot be much strengthened. As noted in Hemaspaandra et al. [HHN+93], Cai et al. [CGH+88] have proven that there is a relativized world in which there are NP-$\leq_T^p$-hard P-selective sets yet the boolean hierarchy does not collapse (so, in particular, $\text{P} \neq \text{NP} \neq \text{coNP}$ in that relativized world). And Kadin [Kad89] has proven the (incomparable) result that for any nice function $f = o(\log n)$ there is a relativized world in which there are NP-$\leq_T^p$-hard P-selective sets yet the polynomial hierarchy does not collapse to $\text{P}^{\text{NP}[f(n)]}$. Also, regarding nonrelativized results related to whether Theorem 4.9 can be strengthened, note that (a) $\text{SAT} \not\in \text{P/poly} \implies \text{EXP} \not\subseteq \text{P/poly}$ (as $\text{SAT} \in \text{EXP}$), and (b) $\text{P} = \text{NP} \implies \text{EXP} \not\subseteq \text{P/poly}$ (by the argument of part 4 of the proof of Theorem 4.8). As (a) and (b) together are logically identical to the statement $[\text{NP} \subseteq \text{R}_T^p(\text{P-sel}) \implies \text{P} = \text{NP}] \implies \text{EXP} \not\subseteq \text{P/poly}$, extending the conclusion of Theorem 4.9 to $\text{P} = \text{NP}$ would have a structural consequence, albeit one that is generally expected to be true. Nonetheless, it is not currently known to be true; though classes are known that are not subsets of P/poly (e.g., as proven by Buhrman, Fortnow, and Thierauf [BFT98], the class known as MA_{EXP}), none of them is currently known to be contained in EXP.

Part 2 of Theorem 4.11 is due to Toda [Tod91]. More on the parallel census technique can be found in the tutorial by Glaßer and Hemaspaandra [GH00]. Part 3 of Theorem 4.11 was proven by Toda [Tod91] and Beigel [Bei88]; see either for a proof of part 3. Sivakumar [Siv99] has shown that Parts 2 and 3 (and a bit more) follow from a hypothesis that is (in light of Tantau's [Tan02] result that $\text{P-mc}(3) - \text{R}_{tt}^p(\text{P-sel}) \neq \emptyset$) weaker than $\text{NP} \subseteq \text{R}_{tt}^p(\text{P-sel})$.

Parts 4 and 5 of Theorem 4.11 are due to Naik and Selman [NS99]. We note in passing that the key idea of the proof of part 4 of Theorem 4.11 doesn't require the full power of that theorem's hypothesis. That is, a stronger result is implicit in the crisp, lovely technique of Naik and Selman. In particular, the proof is really merely using the hypothesis in order to ensure that the $\text{FP}^{\text{NP}}_{tt}$ function has P-enumerable (in the sense of Cai and Hemaspaandra [CH89,CH91]; see below) pronouncements (in the sense of Hemaspaandra and Wechsung [HW91], except now for the case of truth-table rather than Turing reductions; see below). In fact, the hypothesis $\text{NP} \subseteq \text{R}^p_{tt}(\text{P-sel})$ implies that *every* "realization" (via a specific polynomial-time truth-table reduction to a specific NP set) of every $\text{FP}^{\text{NP}}_{tt}$ function has P-enumerable pronouncements. But all we actually need is the potentially weaker assumption that each $\text{FP}^{\text{NP}}_{tt}$ function has *some* realization that has P-enumerable pronouncements. We state this more formally as follows.

Definition 4.23

1. *[CH89] A function $f : \Sigma^* \to \Sigma^*$ is P-enumerable exactly if there is a polynomial-time function that on each input x outputs a collection of strings at least one of which is $f(x)$.*
2. *([HW91], modified to the case of truth-table reductions) For any polynomial-time function-computing deterministic machine M and any NP set A that M accesses in a truth-table fashion to compute its function, the* pronouncement function *of M and A is the function that on each input x is the bit-vector of answers to the queries $M^A(x)$ makes to A.*

Theorem 4.24 *Any function $f \in \text{FP}^{\text{NP}}_{tt}$ that has some realization (a polynomial-time deterministic machine M and an oracle $A \in \text{NP}$ that M accesses in a truth-table fashion, such that M^A computes f) having a P-enumerable pronouncement function is in the class $\text{FP}^{\text{NP}[O(\log n)]}$.*

Corollary 4.25 *If each $\text{FP}^{\text{NP}}_{tt}$ function has some realization having a P-enumerable pronouncement function, then $\text{FP}^{\text{NP}}_{tt} \subseteq \text{FP}^{\text{NP}[O(\log n)]}$.*

The proof of Theorem 4.24 is just like the proof of part 4 of Theorem 4.11, except with the appropriate substitution of our weakened hypothesis, which suffices.

The $k = 1$ special case of Theorem 4.12 was obtained by Hemaspaandra et al. [HHO+93] and Buhrman and Torenvliet [BT96], and the general case ($k \geq 1$) was obtained by Agrawal and Arvind [AA96], Beigel, Kummer, and Stephan [BKS95a], and Ogihara [Ogi95]. The fact that this result applies even for NP-$\leq^p_{O(n^{1-\epsilon})-tt}$-hardness was first noted by Ogihara [Ogi94, Ogi95]. The pruning proof we give of Theorem 4.12 is of a "bottom-up" flavor. Sivakumar [Siv98] has interestingly observed that one can alternatively prove this result in a "top-down" fashion. Lemma 4.13 is due to Jenner and Torán [JT95].

Parts c, d, e, h, and i of Theorem 4.14 are due to Hemaspaandra et al. [HNOS96b], and parts a, b, f, and g of Theorem 4.14 are due to

Hemaspaandra et al. [HHN+95]. Theorems 4.15, 4.17, and 4.19 are due to Hemaspaandra et al. [HNOS96b]. Theorem 4.20 and Corollary 4.22 are due to Hemaspaandra et al. [HHN+95].

Finally, using the recent results "$NP \subseteq P/poly \implies PH = S_2$" [Cai01] and "$NP \subseteq (NP \cap coNP)/poly \implies PH = S_2^{NP \cap coNP}$" [CCHO01], all the theorems of this chapter having the conclusion $PH = ZPP^{NP}$ in fact can be shown in stronger form, namely, with a $PH = S_2$ conclusion when $NP \subseteq P/poly$ is the source of the collapse, e.g., Theorem 4.9 and part 1 of Theorem 4.11, and with a $PH = S_2^{NP \cap coNP}$ conclusion when $NP \subseteq (NP \cap coNP)/poly$ is the source of the collapse, e.g., Corollary 4.10 and part e of Theorem 4.14.

5. Closures

In this chapter we will study several closure properties of the P-selective sets. That is, for various functions h we will ask whether $h(A_1, \ldots, A_k)$ is P-selective whenever all the A_i's are P-selective. Theorem 1.8 states that for complementation (that is, $k = 1$ and $h(A_1) = \overline{A_1}$) the answer is yes; the complement of a P-selective set is always P-selective. In this chapter we will be in part concerned with closure under boolean operations, i.e., with functions of the form $\chi_{h(A_1,\ldots,A_k)} = f(\chi_{A_1}, \ldots, \chi_{A_k})$ for some boolean function f. In particular, we will study so-called boolean *connectives*. A boolean connective $I(f)$ is defined using a boolean operator f; $I(f)$ works on a collection of sets and returns a single set using, in the fashion just described, the boolean operator f with which it is defined. For instance, let f_{or} denote the boolean function such that $f_{or}(a, b) = a \vee b$. The connective $I(f_{or})$ works on two sets A and B and, since it applies an element-by-element "or," returns $A \cup B$ (i.e., $I(f_{or})(A, B) = A \cup B$). Indeed, a boolean connective $I(f)$ can be defined for any boolean function f. Since there are 2^{2^k} boolean functions on k variables, it follows that there are 2^{2^k} potential boolean closures to be investigated for the P-selective sets. Section 5.2 will give a complete account of this investigation. In particular, for each k we will determine under exactly how many—and which—k-ary boolean connectives the P-selective sets are closed. We will see that of the 2^{2^k} k-ary boolean functions, exactly $2k + 2$ are functions defining boolean connectives under which the P-selective sets are closed.

Another well-studied subject is the (downward) closure of P-selective sets under various types of reductions. Many important complexity classes are closed downward under a wide variety of polynomial-time reductions. For example, NP is closed under many-one reductions, conjunctive reductions, and disjunctive reductions. EXP is even closed under Turing reductions. P-sel is closed under positive Turing reductions, as we will see in Section 5.3. However, as Section 5.5 will establish, P-sel is not closed under even the weakest of nonpositive reductions.

In Section 5.4 we study the combination of self-reducibility and selectivity. The P-selective self-reducible sets are closed downward under 1-truth-table reductions. We prove this via the stronger result that any self-reducible set that 1-truth-table reduces to a P-selective set necessarily belongs to P. We prove analogous results relating self-reducibility and nondeterministic selectivity to membership in nondeterministic classes.

5.1 Connectives and Reductions

We formally define two notions that will play an important role in this chapter: *boolean connectives* and *positive reductions*. We will need boolean connectives in Section 5.2, where we discuss whether P-selective sets of various types are closed under these connectives, and we will need positive reductions in Section 5.3, where we discuss reduction closures.

Definition 5.1   *Let f be a k-ary boolean function. Define $I(f)$ by*

$$I(f)(A_1, \ldots, A_k) = \{x \in \Sigma^* \mid f(\chi_{A_1}(x), \ldots, \chi_{A_k}(x))\}.$$

We call $I(f)$ a k-ary boolean connective.

Some particularly simple connectives will play a central role in our results. We will call a k-ary connective *completely degenerate* if its associated boolean function is a constant function. We will call a connective *almost-completely degenerate* if the result of its boolean function is dependent on at most one variable. That is, if either it is a 0-ary connective or it has a variable such that if we fix that variable to a "true" then the resulting connective is completely degenerate and if we fix that variable to "false" then the resulting connective is also completely degenerate. Note that all completely degenerate connectives clearly are also almost-completely degenerate. We will also speak of degenerate, completely degenerate, and almost completely degenerate *functions*, meaning that the connective corresponding to the function is degenerate, completely degenerate, or almost completely degenerate, respectively.

The well-known boolean functions $\wedge$ and $\vee$ (viewed as 2-ary functions rather than infix operators) give, respectively, the boolean connectives $\cap$ and $\cup$, e.g., $I(\wedge)(A_1, A_2) = \{x \in \Sigma^* \mid x \in A_1 \wedge x \in A_2\} = A_1 \cap A_2$.

In Section 5.3 we will discuss the (downward) closure of selective sets of various kinds under different types of reduction. One type of reduction under which the P-selective sets are closed is the so-called positive (Turing) reduction. Positivity is a restriction on the reduction types we have already seen. Almost any kind of reduction can be restricted to being positive as follows. We demand that, for some machine realizing the reduction, adding more strings to any oracle set being accessed does not change the outcome of the reduction from accepting to rejecting. There are two variants of positive reductions: locally and globally positive reductions. The notion described above is global positivity, and the results in this chapter are stated for globally positive reductions. Globally positive reductions are defined as follows (see also Appendix A.1).

Definition 5.2 $A \leq^p_{pos} B$ *if there is a deterministic polynomial-time Turing machine M such that*

1. $A = L(M^B)$, *and*
2. $(\forall C, D) [C \subseteq D \implies L(M^C) \subseteq L(M^D)]$.

Examples of positive reductions are many-one reductions, conjunctive reductions, and disjunctive reductions. An example of a nonpositive reduction is the exclusive-or (parity) reduction—the reduction that asks two queries to the oracle and accepts if and only if the number of yes answers to these queries is odd.

5.2 Boolean Closures

In this section we will determine exactly which connectives the class of P-selective sets is closed under. That is, we study those connectives that when applied to P-selective sets always result in a P-selective set. Note that taking the boolean closure of a complexity class slightly resembles in flavor, but is not necessarily the same as, taking the closure of that class under bounded-truth-table reductions. Though queries in a bounded truth-table are evaluated by a boolean function, these queries are evaluated with respect to the characteristic function of a *single* P-selective set. In contrast, a boolean connective works on the characteristic functions of a *collection* of sets. Phrased differently, in a bounded-truth-table reduction you compose the different characteristic values of *different words* with respect to a *single set*, whereas in a boolean connective you compose the characteristic values of a *single word* with respect to *different sets*.

As noted earlier, P-sel is closed under complementation, by which we formally mean $(\forall A \in \text{P-sel})[\overline{A} \in \text{P-sel}]$. This is also true for some nondeterministic selectivity classes.

Theorem 5.3 NPSV_t-sel *and* NPMV_t-sel *are closed under complementation.*

Proof Given a selector function $f \in \text{NPSV}_t$, NPSV_t clearly also contains the function $f'(x, y)$ that is $\{x, y\} - f(x, y)$ when $x \neq y$ and is x when $x = y$. Moreover, if f is a selector for A then f' is a selector for $\overline{A}$.

If M is an NP machine that computes an NPMV_t selector function f for a set A, then an NPMV_t selector for $\overline{A}$ is computed by the machine M' that on input (x, y) simulates M on (x, y), except on each simulated path it outputs x if M would output y on that path and outputs y if M would output x on that path. □

Also, we have the following equivalences.

Theorem 5.4 *The following are equivalent:*

1. NPMV-sel *is closed under complementation.*
2. $\text{NPMV}_t\text{-sel} = \text{NPMV-sel}$.
3. $\text{NP} = \text{coNP}$.

Proof Parts 1 and 2 are clearly equivalent in light of Theorem 1.28 and Theorem 5.3. If NPMV-sel is closed under complementation, then in light of Proposition 1.25 it holds that $\overline{\text{SAT}} \in \text{NPMV-sel}$, and thus $\text{coNP} \subseteq \text{NPMV-sel}$. So by Theorem 4.14 we have $\text{NP} = \text{coNP}$. Finally, if $\text{NP} = \text{coNP}$, then $\text{NPMV}_t\text{-sel} = \text{NPMV-sel}$ by Proposition 1.32. $\square$

Next, by showing the existence of two P-selective sets L and R such that $L \cap R$ is not P-selective, we will prove that P-sel is not closed under intersection. Theorem 5.5 will capture the nonclosure under intersection simultaneously for the main types of selectivity that are of interest to us in this book (see Corollary 5.6). The fact that none of these classes are closed under intersection will then be obtained as a corollary. The theorem will also have as a corollary that our selectivity classes are not closed under union, as stated in Corollary 5.7.

In Chapter 1, we introduced the standard left cuts of real numbers as an example of P-selective sets. In the next theorem we will use a variation on this theme. Instead of including all strings that are lexicographically less than a real number, we include all the strings that are lexicographically less than *or equal* to the *prefix* (that is the same length as the string) of an infinite bitstring. The set L used in the following proof is such a set, and the set R is the "right" (as opposed to "left") analog of such a set. Clearly both these types of sets are P-selective.

Theorem 5.5 *There exist P-selective sets whose intersection is not P-selective (and indeed not even semi-recursive).*

Proof Our P-selective sets whose intersection is not semi-recursive will be called L and R. In the following, a *branch* B is simply an infinite string of bits, that is, $B = b_0 b_1 b_2 \cdots$. We associate a language L_B with B as follows: $L_B = \{b \mid b \text{ is a prefix of } B\}$. If we define $L = \{b \mid b \leq_{\text{lex}} b_0 b_1 \cdots b_{|b|}\}$ and $R = \{b \mid b \geq_{\text{lex}} b_0 b_1 \cdots b_{|b|}\}$, then $L \cap R = L_B$. It remains to choose B such that L_B is not semi-recursive. We use a structural diagonalization. Let $\widehat{M}_0, \widehat{M}_1, \ldots$ be an enumeration of all machines that could possibly witness that L_B is semi-recursive. Let B already be constructed up to the ith level. We now have to choose $b_{i+1} \in \{0, 1\}$. If $\widehat{M}_i$ on input $(b_0 \cdots b_i 0, b_0 \cdots b_i 1)$ halts and chooses $b_0 \cdots b_i 0$ let $b_{i+1} = 1$, otherwise let $b_{i+1} = 0$. Clearly $\widehat{M}_i$ is now guaranteed to err on input $(b_0 \cdots b_i 0, b_0 \cdots b_i 1)$. $\square$

Now we can conclude the nonclosure under intersection for all selectivity classes of interest in this book, as is stated in the following corollary, which follows due to the easy-to-see fact that $\text{NPMV-sel} \subseteq \text{FEXP-sel}$ where FEXP denotes the single-valued total functions computable in $\bigcup_{k>0} \text{DTIME}[2^{n^k}]$.

Corollary 5.6 *The classes* P-sel, NPSV$_t$-sel, NPSV-sel, NPMV$_t$-sel, *and* NPMV-sel *are not closed under intersection.*

The P-selective sets are closed under complementation (Theorem 1.8). So the complements of the P-selective sets L and R in the proof of Theorem 5.5 are also P-selective sets. Also, the semi-recursive sets are closed under complementation and so, equivalently, the class of all non-semi-recursive sets is closed under complementation. Thus, we obtain the following corollary.

Corollary 5.7 *Consider the following operations on languages X and Y:* $X \cap Y$, $\overline{X} \cap \overline{Y}$, $\overline{X} \cap \overline{Y}$, $X \cap \overline{Y}$, $X \cup Y$, $\overline{X} \cup Y$, $X \cup \overline{Y}$, $\overline{X} \cup \overline{Y}$. *In each of these cases P-selective sets X and Y can be found such that the language resulting from the operation is not semi-recursive (and so is not in the classes* NPMV-sel, P-sel, NPSV$_t$-sel, NPSV-sel, *and* NPMV$_t$-sel*).*

The proof of Theorem 5.6 has another interesting corollary. As the intersection of the two P-selective sets created in the proof of that theorem consists of a single string at each length and, moreover, at these lengths the intersection of the *complements* of these sets is empty, we can conclude the following theorem for the remaining two nondegenerate connectives. (Note: A NXOR B denotes $(A \cap B) \cup (\overline{A} \cap \overline{B})$.)

Theorem 5.8 *There are* P-*selective sets X and Y such that $(X \cap Y) \cup (\overline{X} \cap \overline{Y})$ and $(X \cap \overline{Y}) \cup (\overline{X} \cap Y)$ are not semi-recursive.*

Proof In the proof of Theorem 5.5, the sets there called L and R are chosen such that $L \cap R$ consists of exactly one string at each length. This means that their complements are disjoint and that L NXOR R equals $L \cap R$. Thus, if we take the same sets L and R, the same proof says that L NXOR R is not semi-recursive. Closure under complement of the semi-recursive sets now gives the second statement of the theorem. □

Corollary 5.9 P-sel, NPSV$_t$-sel, NPSV-sel, NPMV$_t$-sel, *and* NPMV-sel *are not closed under* NXOR *and* XOR *connectives.*

For 2-ary connectives we now have counted ten connectives (namely, the ten nondegenerate 2-ary connectives) under which the P-selective sets are not closed. Two 2-ary connectives are completely degenerate (the connectives corresponding to the boolean functions TRUE and FALSE) and four 2-ary connectives are almost-completely degenerate, the ones corresponding to the boolean functions $f(x, y) = x$, $f(x, y) = y$, $f(x, y) = \neg x$, and $f(x, y) = \neg y$. In a more sloppy notation, below, we will also sometimes use for these four connectives the characteristic functions of the resulting set, i.e., with respect to being applied to some sets A and B the four just-mentioned connectives might (in this sloppy shorthand) be written, respectively: $\chi(A)$, $\chi(B)$, $\neg \chi(A)$, and $\neg \chi(B)$. Of course P-sel, NPSV$_t$-sel, and NPMV$_t$-sel are all closed under these connectives. The two remaining classes NPSV-sel and NPMV-sel are closed under all except possibly $\neg\chi(A)$, and $\neg\chi(B)$ (see Theorem 5.4 regarding complementation and NPMV-sel, and note that it does not address NPSV-sel). So, in light of these comments, Corollary 5.9, and Corollary 5.7, the degenerate 2-ary connectives are the only connectives under which the classes P-sel, $NPSV_t$-sel, and $NPMV_t$-sel are closed. Summing up, we have the following theorem.

Theorem 5.10 *Regarding 2-ary boolean connectives, the situation is as follows. The classes P-sel, $NPSV_t$-sel, and $NPMV_t$-sel are closed under exactly the degenerate 2-ary connectives. The classes NPSV-sel and NPMV-sel are not closed under nondegenerate 2-ary connectives.*

We will now generalize this statement to k-ary connectives.

Theorem 5.11 *For any $k \geq 0$, regarding k-ary boolean connectives, the classes P-sel, $NPSV_t$-sel, and $NPMV_t$-sel are closed under exactly the degenerate k-ary connectives. The classes NPSV-sel and NPMV-sel are not closed under nondegenerate k-ary connectives.*

Proof The $k = 0$ case is immediate. In the case $k = 1$ there are four connectives: identity, complementation, and the completely degenerate connectives that yield $\emptyset$ and Σ^* respectively. All are degenerate and P-sel, $NPSV_t$-sel, and $NPMV_t$-sel are closed under each of these. The case $k = 2$ appears as Theorem 5.10. In fact, for $k > 2$ it is obvious, for the same reason as in our earlier arguments, that all $2k+2$ degenerate connectives are closure properties of the classes P-sel, $NPSV_t$-sel, and $NPMV_t$-sel.

All that remains to be shown is that, for $k > 2$, all nondegenerate connectives are not closure properties of any of the classes P-sel, $NPSV_t$-sel, NPSV-sel, $NPMV_t$-sel and $NPMV_t$-sel. We will prove this by induction, with the basis cases ($k = 1$ and $k = 2$) already having been done above.

Suppose we have proven the theorem for all k-ary connectives. Let f be a $(k+1)$-ary nondegenerate boolean function on the variables $x_1, x_2, \ldots, x_{k+1}$. The variables correspond to the characteristic functions of the languages $L_1, L_2, \ldots, L_{k+1}$. Let $f_{x_i=b}$ stand for the function on k variables obtained by substituting the value b for the variable x_i. If either $f_{x_1=0}$ (respectively, $f_{x_1=1}$) is nondegenerate we fix L_1 to $\emptyset$ (respectively, Σ^*) and we have the nonclosure of P-sel under f by the inductive hypothesis. So we may assume that both k-ary boolean functions, $f_{x_1=0}$ and $f_{x_1=1}$, are degenerate.

This means that $f_{x_1=0}$ and $f_{x_1=1}$ can either be identically 0, identically 1, or for some $i > 1$ behave exactly as χ_{L_i}, or $\chi_{\overline{L_i}}$. In the following table we state all possible cases. The table omits the cases where the roles of L_i and L_j are interchanged and where $\overline{L_i}$ can be taken in the place of L_i.

$f_{x_1=0}$	$f_{x_1=1}$	f	which is
0	0	0	degenerate
0	1	χ_{L_1}	degenerate
1	0	$\chi_{\overline{L_1}}$	degenerate
1	1	1	degenerate
0	χ_{L_i}	$\chi_{(L_1 \cap L_i)}$	$\chi_{(L_1 \cap L_i)}$
1	χ_{L_i}	$\chi_{(\overline{L_1} \cup (L_1 \cap L_i))}$	$\chi_{(\overline{L_1} \cup L_i)}$
χ_{L_i}	χ_{L_j}	$\chi_{(\overline{L_1} \cap L_i)} \cup \chi_{(L_1 \cap L_j)}$	$\chi_{(L_1 \cap L_j)}$, if we let $L_1 = L_i$

For cases in this table in which f is nondegenerate, it follows from Corollary 5.7 that we can choose P-selective sets L_i and L_j such that $I(f)(L_i, L_j)$ is not semi-recursive, where I is as per Definition 5.1.   □

5.3 Reductions Under Which P-sel Is Closed Downward

In this section we will prove the closure of P-sel under certain types of reductions, i.e., we will show that for certain reductions r it holds that if $A \leq_r B$ and B is a P-selective set, then so is A.

In Chapter 1 we saw that P-sel is closed under many-one reductions (Theorem 1.5). However, there are forms of reduction under which P-sel is not closed. For example, we noted in Chapter 1 that any tally set is Turing reducible to a P-selective set, and yet there are tally sets that are not P-selective. In fact, as we will see in Section 5.5, P-sel is not closed under any form of nonpositive reduction. The following theorem states a closure result for positive reductions.

Theorem 5.12 *For any set A and any P-selective set B, both neither empty nor equal to Σ^*, if $A \leq_{pos}^p B$ then $A \leq_m^p B$.*

Corollary 5.13 *For any set A and any P-selective set B, if $A \leq_{pos}^p B$ then A is P-selective.*

Before we prove the theorem, we will prove a lemma that makes the same statement for positive-truth-table reductions. The proof of the theorem then consists of converting a positive Turing reduction to a P-selective set to a positive-truth-table reduction to a P-selective set.

Lemma 5.14 *For any set A and any P-selective set B, both neither empty nor equal to Σ^*, if $A \leq_{ptt}^p B$ then $A \leq_m^p B$.*

Proof Let f be a P-selector for B. Let $A \leq_{ptt}^p B$ via machine M that on input x produces the queries $q_1, \ldots, q_{p(|x|)}$ and without loss of generality (via using the function $order$ of Lemma 4.6), let it be the case that $q_1, \ldots, q_{p(|x|)}$ have the property that the actual list of membership answers in B are guaranteed to be of the form $0^j 1^{p(|x|)-j}$ for some $0 \leq j \leq p(|x|)$. If M either accepts when simulated in every one of these cases or rejects when simulated in every one of these cases, then we are done. In such a case, we have the reduction output a fixed string in B if M always accepted and we have it output a fixed string in $\overline{B}$ if M always rejected. Otherwise, keeping in mind that M is implementing a *positive*-truth-table reduction, there will be a j, $1 \le j \le p(|x|)$, such that M will reject in the simulation with $0^j 1^{p(|x|)-j}$ and accept in the simulation with $0^{j-1} 1^{p(|x|)-j+1}$. In this case M accepts x if and only if $q_j \in B$. So we complete the specification of our many-one reduction by in this case making our output be precisely this q_j. $\quad\square$

Before we show how to convert a positive Turing reduction into a positive-truth-table reduction, let us present a lemma that in its statement also introduces some notation.

Lemma 5.15 *Let B be a P-selective set with symmetric selector f, let $B^+(z) = \{x \mid f(x, z) = x\}$, and let $B^-(z) = B^+(z) - \{z\}$. It holds that*

1. $z \in B \implies B^+(z) \subseteq B$, and
2. $z \notin B \implies B \subseteq B^-(z)$.

Proof 1. If $x \in B^+(z)$ then $f(x, z) = x$, so $z \in B \implies x \in B$.
2. If $x \in B$ and $z \notin B$ then $x \ne z$ and $f(x, z) = x$, so $x \in B^-(z)$. $\quad\square$

Proof of Theorem 5.12   Let B be a P-selective set and let $A \le_{pos}^p B$. We show that $A \le_{ptt}^p B$. Let M be an oracle machine witnessing $A \le_{pos}^p B$, and suppose that M is time-bounded by some polynomial p. On input x the following algorithm produces a list of queries $q_1, \ldots, q_k$, where $k \le p(|x|)$, and a positive truth-table α such that $x \in A$ if and only if $\alpha(\chi_B(q_1), \chi_B(q_2), \ldots, \chi_B(q_k))$ is true.

The first query in the list q_1 can be obtained by running $M^\emptyset(x)$ for at most $p(|x|)$ steps. If $M^\emptyset(x)$ halts without producing a query, then it either accepts or rejects. The output q_1 is then a fixed string in B if $M^\emptyset(x)$ accepts or a fixed string in $\overline{B}$ if $M^\emptyset(x)$ rejects, and the truth-table is just $\chi_B(q_1)$. This is in fact then already a many-one reduction. So assume that $M^\emptyset(x)$ produces at least one query, q_1.

Suppose that we have a list of queries $q_1, \ldots, q_i$ and a list of answers $a_1, \ldots, a_{i-1}$, where $i \ge 1$, with the property that, if M is simulated on input x and with $a_1, \ldots, a_{i-1}$ in the place of oracle answers, then $q_1, \ldots, q_i$ is the list of queries produced by M. The sequence $a_1, \ldots, a_{i-1}$ will be a list of bits, where 1 is identified with a yes answer and 0 is identified with a no answer; this makes it easier to identify these answers with values of χ_B and truth values of the truth-table. We have assumed that this statement is true for $i = 1$, where q_1 is the first query produced and the list of answers is empty.

Consider the computations $M^{B^+(q_i)}(x)$ and $M^{B^-(q_i)}(x)$. There are four possible outcomes for this pair of computations. However, since M is a positive reduction, only three of these are valid.

1. $M^{B^+(q_i)}(x)$ and $M^{B^-(q_i)}(x)$ both reject.

2. $M^{B^+(q_i)}(x)$ and $M^{B^-(q_i)}(x)$ both accept.

3. $M^{B^+(q_i)}(x)$ accepts and $M^{B^-(q_i)}(x)$ rejects.

If case 3 occurs, we on this input obtain a positive-truth-table (even many-one) behavior from the simulation. The truth-table behavior in this case is simply querying q_i and accepting if and only if $\chi_B(q_i) = 1$, since in this case $x \in A$ if and only if $q_i \in B$. This can easily be seen to be correct by observing the following implications:

1. $q_i \in B \implies B^+(q_i) \subseteq B$.
   $[B^+(q_i) \subseteq B \wedge M^{B^+(q_i)}(x) \text{ accepts}] \implies M^B(x) \text{ accepts}$.
   $M^B(x) \text{ accepts} \implies x \in A$.
2. $q_i \notin B \implies B \subseteq B^-(q_i)$.
   $[B \subseteq B^-(q_i) \wedge M^{B^-(q_i)}(x) \text{ rejects}] \implies M^B(x) \text{ rejects}$.
   $M^B(x) \text{ rejects} \implies x \notin A$.

So, we are left with these two cases:

1. $M^{B^+(q_i)}(x)$ and $M^{B^-(q_i)}(x)$ both reject.
2. $M^{B^+(q_i)}(x)$ and $M^{B^-(q_i)}(x)$ both accept.

In both these cases we produce a next answer a_i in the list, namely the answer yes in case 1 and the answer no in case 2. By simulating M on input x we assumed that the list $q_1, \ldots, q_i$ of queries was generated when the list $a_1, \ldots, a_{i-1}$ was used instead of oracle answers. Since we now have decided on an answer a_i, we can continue the simulation and either produce a query q_{i+1} or find that the simulation ends without producing more queries. Since the simulation thus described follows a single computation path of M on input x, the simulation will halt after at most $p(|x|)$ simulated computation steps and after producing $q_1, \ldots, q_k$, where $k \leq p(|x|)$. Then it accepts or rejects; to the list of answers we add an extra bit a_{k+1} which is 1 if this simulation ends in acceptance and is 0 if this simulation ends in rejection. We must now define a positive truth-table $\alpha(\chi_B(q_1), \ldots, \chi_B(q_k))$. This is done as follows.

1. If $\chi_B(q_j) = a_j$ for $j \leq k$ then $\alpha(\chi_B(q_1), \ldots, \chi_B(q_k)) = a_{k+1}$.
2. If $(\exists j \leq k)[\chi_B(q_j) < a_k]$ then $\alpha(\chi_B(q_1), \ldots, \chi_B(q_k)) = 0$.
3. If $(\exists j \leq k)[\chi_B(q_j) > a_k]$ then $\alpha(\chi_B(q_1), \ldots, \chi_B(q_k)) = 1$.

Note that we have presented the truth-table here in the "query generation" and "answer interpretation" style (see also Appendix A.1) to make the positive nature of the truth-table reduction even more clear. When we picture the possible oracle answers to the queries as a string, the cases where the truth-table reduction accepts x are all lexicographically less than the cases where the truth-table reduction accepts x. We must now argue that it is also correct, i.e., $x \in A$ if and only if $\alpha(\chi_B(q_1), \ldots, \chi_B(q_k)) = 1$. Case 1 is easy. In case 2, let q_j be the first such query in the list, i.e., $\chi_B(q_j) = 0$ and $a_j = 1$. From $a_j = 1$ we have that $M^{B^-(q_j)}(x)$ rejects and from $\chi_B(q_j) = 0$ we have that $B \subseteq B^-(q_i)$, so that $M^B(x)$ rejects and $x \notin A$. In case 3, let q_j be the first such query in the list, i.e., $\chi_B(q_j) = 1$ and $a_j = 0$. From $a_j = 0$ we have that $M^{B^+(q_i)}(x)$ accepts and from $\chi_B(q_j) = 1$ we have that $B^+(q_i) \subseteq B$, so that $M^B(x)$ accepts and $x \in A$. We note finally that case 2 and case 3 cannot occur simultaneously since that would imply a contradiction. This means that, always, exactly one of the three cases occurs, and thus that the reduction is well-defined.   ❑   Theorem 5.12

A result similar to Theorem 5.12 can be obtained for reductions to NPSV_t-selective sets. The following lemma is obtained by relativization of Theorem 5.12 to an oracle L.

Lemma 5.16   *If $A \leq_{pos}^p B$, B is FP_t^L-selective for some set L, $B \neq \emptyset$, and $B \neq \Sigma^*$, then $A \leq_m^{p,L} B$ and hence A is FP_t^L-selective.*

Theorem 5.17   *If $A \leq_{pos}^p B$ and B is NPSV_t-selective then*

1. *A is NPSV_t-selective, and*
2. *if $B \neq \Sigma^*$ and $B \neq \emptyset$ then $A \leq_m^{\text{NPSV}_t} B$.*

Proof   Let B be NPSV_t-selective. There exists a language L in $\text{NP} \cap \text{coNP}$ such that B is FP_t^L-selective. Since $L \in \text{NP} \cap \text{coNP}$ it follows from Lemma 1.30 that $A \leq_m^{\text{NPSV}_t} B$ and that A is NPSV_t-selective.   ❑

5.4 Self-Reducible Sets and Selectivity

For the rest of this chapter, when f is a symmetric P-selector function, $a \leq_f b$ denotes $f(a, b) = b$.

Self-reducibility was defined in Chapter 1. Recall that a set A is Turing self-reducible if there is a deterministic polynomial-time Turing machine M such that $A = L(M^A)$ and for each x it holds that $M^A(x)$ queries strings only of lengths strictly less than $|x|$. In this section we will see that the only sets that are both Turing self-reducible and P-selective are the polynomial-time computable sets. Strictly speaking, this is of course not a reduction closure. However, the theorem we state says that any set that is Turing self-reducible and 1-truth-table *reducible* to a P-selective set is in P (from which it follows that it is P-selective and self-reducible). Thus, it follows from this theorem that the intersection of the class P-sel and the class of Turing self-reducible sets is closed downward under 1-truth-table reductions. In fact, this is a little bit more general since the P-selective set reduced *to* need not be self-reducible. Since Turing self-reducible sets are all in PSPACE, the following theorem is somewhat related to part 3 of Theorem 4.8.

Theorem 5.18   *Let A be Turing self-reducible and let B be P-selective. If $A \leq_{1\text{-}tt}^p B$ then $A \in \text{P}$.*

Proof Let A be Turing self-reducible via M. Let A be 1-truth-table re-ducible to B via M_r. Let f be a symmetric (via Theorem 1.4) P-selector for B. For x in Σ^*, $q(x) \in \Sigma^*$ and $v(x) \in \{0,1\}$ will be such that M_r accepts x if and only if $\chi_B(q(x)) = v(x)$. In particular, on input x, let $q(x)$ be the single query that M_r asks of B, and define $v(x)$ to capture $M_r(x)$'s accep-tance behavior, i.e., $M_r(x)$ accepts if and only if the answer is yes in the case $v(x) = 1$, and accepts if and only if the answer is no in the case $v(x) = 0$. Note that $y \leq_f z$ implies $y \in B \implies z \in B$. Let $\perp$ and $\top$ be new symbols such that $\perp \leq_f y \leq_f \top$ for every $y \in \Sigma^*$.

Let x be some input and let $q(x)$ and $v(x)$ be as above. We will give a polynomial-time algorithm that decides $x \in A$.

Consider a simulation M_s of M on x defined as follows. The simulation will use two strings t and b. Initially t is set to $\top$ and b is set to $\perp$. Set $i = j = 0$. M_s simulates M such that when M makes the kth query, y_k, then M_s performs the following steps:

1. Simulate M_r on y_k to compute $q(y_k)$ and $v(y_k)$.
2. If $t \leq_f q(y_k)$ then choose the branch corresponding to $q(y_k) \in B$, i.e., choose the branch corresponding to $y_k \in A$ if $v(y_k) = 1$ and the one corresponding to $y_k \notin A$ if $v(y_k) = 0$. If $q(y_k) \leq_f b$ then choose the branch corresponding to $q(y_k) \notin B$. If $b \leq_f q(y_k) \leq_f t$ then compute $f(q(x), q(y_k))$. If $q(x) \leq_f q(y_k)$ then set t to $q(y_k)$ and choose the branch corresponding to $q(y_k) \in B$. If $q(y_k) \leq_f q(x)$ then set b to $q(y_k)$ and choose the branch corresponding to $q(y_k) \notin B$.

Let r be 1 if M_s accepts in the simulation and let r be 0 otherwise. Let t_0 and b_0 be the final values of t and b, respectively, and let i and j be such that t_0 is set to $q(y_i)$ and b_0 is set to $q(y_j)$. It may of course happen that t or b do not get new values in the course of this simulation. If, for example, t does not get a new value, $t_0 = \top$ and $i = 0$. The following hold:

1. $b_0 \leq_f q(x) \leq_f t_0$.
2. If $b_0 \notin B$ and $t_0 \in B$ then: M^A on x accepts if and only if $r = 1$. So in this case $\chi_A(x) = r$.
3. If $b_0 \in B$ then $q(x) \in B$. So in this case $\chi_A(x) = v(x)$.
4. If $t_0 \notin B$ then $q(x) \notin B$. So in this case $\chi_A(x) = 1 - v(x)$.

Suppose $r = v(x)$. Then we have that $\chi_A(x) = r$ if and only if $t_0 \in B$. (Note that $i = 0$ implies $t_0 \in B$ and thus $\chi_A(x) = r$.) So $\chi_A(x) = r$ if and only if $q(y_i) \in B$, which itself holds if and only if $\chi_A(y_i) = v(y_i)$. If this case holds (i.e., $r = v(x)$): set $z = y_i$, set $e = 1$ if $r = v(y_i)$, and set $e = 0$ if $r \neq v(y_i)$.

Suppose $r = 1 - v(x)$. Then we have $\chi_A(x) = r$ if and only if $b_0 \notin B$. (Note that in this case $j = 0$ implies $b_0 \notin B$ and thus $\chi_A(x) = r$.) So $\chi_A(x) = r$ if and only if $q_j \notin B$, and $q_j \notin B$ if and only if $\chi_A(y_j) = 1 - v(y_j)$. If this case (i.e., $r = 1 - v(x)$) holds: set $z = y_i$, set $e = 0$ if $r = v(y_j)$, and set $e = 1$ if $r \neq v(y_j)$.

In both cases $x \in A$ if and only if $\chi_A(z) = e$. Thus, M_s will find a string z and a value e such that $\chi_A(x) = 1$ if and only if $\chi_A(z) = e$. Moreover, z is in the self-reduction tree of M on input x. This means that $|z| < |x|$. We can use the same simulation to determine the value of $\chi_A(z)$, which may depend on a new string z' with $|z'| < |z|$ and so on. Eventually, after at most $|x|$ simulations, we can just compute the value of χ_A for the string in question, and from that, via the chain of implications built up during these simulations, can compute the value of $\chi_A(x)$ ▫

We will now discuss the self-reducibility results that hold for nondeterministically selective sets. Some basic results can be immediately obtained by relativizing the results we have already obtained. For example, it is easy to see that by carefully relativizing Theorem 5.18 we immediately obtain the following result.

Corollary 5.19 *If A is Turing self-reducible, B is NPSV_t-selective, and $A \leq^p_{1\text{-}tt} B$, then $A \in \mathrm{NP} \cap \mathrm{coNP}$.*

However, the NPSV_t-selectivity here is needlessly restrictive. In fact, as we will now show, the result holds even for NPMV_t-selectivity. Indeed, we will prove a bit more by proving Theorem 5.20, whose proof we deferred earlier. Namely, this result already appeared, without proof, as Theorem 4.20. The definition of γ reductions is given in the appendix and will be important here. The fact that γ reductions come into play here is not surprising. Speaking very informally, moving from deterministic selectivity to NPSV_t-selectivity is like relativizing by $\mathrm{NP} \cap \mathrm{coNP}$, and γ reductions themselves are an "$\mathrm{NP} \cap \mathrm{coNP}$-like" reduction type. The structure of the proof of Theorem 5.20 very much resembles the proof of Theorem 5.18, though of course the nondeterministic nature (of the γ reduction and of the selector function) requires additional care.

Theorem 5.20 *If A is Turing self-reducible and $A \in \mathrm{R}_\gamma(\mathrm{R}^p_{1\text{-}tt}(\mathrm{NPMV}_t\text{-sel}))$ then $A \in \mathrm{NP} \cap \mathrm{coNP}$.*

Proof Let A be Turing self-reducible via M. Let A γ-reduce, via nondeterministic machine N, to a set C that itself is 1-truth-table reducible via M_r to a set B in NPMV_t-sel. We follow the proof structure of Theorem 5.18. For each $x \in \Sigma^*$, for each accepting path p of N on input x we denote by $u(p, x)$ the output of p. Let $q(u(p, x))$ be the string that is queried by M_r on input $u(p, x)$. Let $v(u(p, x)) \in \{0, 1\}$ be such that $x \in A$ if and only if $\chi_B(q(u(p, x))) = v(u(p, x))$. Let f be a symmetric NPMV_t-selector function for B (we may assume symmetry by Theorem 1.24). We use the notation $y \leq_f z$ for $z \in \mathrm{set}\text{-}f(y, z)$. Note that $y \leq_f z$ implies: $y \in B \implies z \in B$. Fix x and let $q_0 = q(u(p, x))$ and $v_0 = v(u(p, x))$.

There exists a nondeterministic polynomial-time algorithm that can prove either $x \in A$ if indeed $x \in A$, or $x \notin A$ if indeed $x \notin A$. The existence of such an algorithm shows that $A \in \mathrm{NP} \cap \mathrm{coNP}$. Consider a simulation (let the machine doing it be called M_s) of M on input x as follows. The simulation will use two strings t and b initially set to $\top$ and $\bot$ respectively, where $\top$ and $\bot$ are new symbols such that $\bot \leq_f y \leq_f \top$ for every $y \in \Sigma^*$, as in the proof of Theorem 5.18. M_s simulates M such that when M makes the kth query y_k, M_s performs the following steps.

1. Simulate N to determine $u(p, y_k)$, $q_k = q(u(p, y_k))$, and $v_k = v(u(p, y_k))$. It now holds that $y_k \in A$ if and only if $v_k = \chi_B(q_k)$.
2. If $t \leq_f q_k$ then choose the branch corresponding to $q_k \in B$.
   If $q_k \leq_f b$ then choose the branch corresponding to $q_k \notin B$.
   If $b \leq_f q_k \leq_f t$ then simulate $f(q_0, q_k)$. If $q_0 \leq_f q_k$ then set t to q_k and choose the branch corresponding to $q_k \in B$. If $q_k \leq_f q_0$ then set b to q_k and choose the branch corresponding to $q_k \notin B$.

Let r be 1 if M_s accepts in this simulation and let r be 0 if M_s rejects. Let t_0 and b_0 be the final values of t and b respectively. Let i and j be such that t_0 is set to q_i and b_0 is set to q_j. The following hold:

1. $b_0 \leq_f q_0 \leq_f t_0$.
2. If $b_0 \notin B$ and $t_0 \in B$ then: M^A accepts if and only if $r = 1$. So in this case $\chi_A(x) = r$.
3. If $b_0 \in B$ then $q_0 \in B$. So in this case $\chi_A(x) = v_0$.
4. If $t_0 \notin B$ then $q_0 \notin B$. So in this case $\chi_A(x) = 1 - v_0$.

Suppose $r = v_0$. Then we have that $\chi_A(x) = r$ if and only if $t_0 \in B$. So $\chi_A(x) = r$ if and only if $q_i \in B$, and $q_i \in B$ if and only if $\chi_A(y_i) = v_i$. In this case (i.e., $r = v_0$): set $z = y_i$, set $e = 1$ if $r = v_i$, and set $e = 0$ if $r \neq v_i$.

Now suppose $r = 1 - v_0$. Then we have that $\chi_A(x) = r$ if and only if $b_0 \notin B$. So $\chi_A(x) = r$ if and only if $q_j \notin B$, which itself holds if and only if $\chi_A(y_j) = 1 - v_j$. If this case (i.e., $r = 1 - v_0$) holds: set $z = y_j$, set $e = 0$ if $r = v_j$, and set $e = 1$ if $r \neq v_j$.

In either case we have $x \in A$ if and only if $\chi_A(z) = e$.

We see that there is a nondeterministic polynomial-time procedure that takes x as input and produces a string z and a bit e such that $x \in A$ if and only if $\chi_A(z) = e$. Note that this is the same as saying $x \notin A$ if and only if $\chi_A(z) = 1 - e$. Moreover, z is in the self-reduction tree for x. So $|z| < |x|$. Recursive application of this simulation therefore gives a list of strings $z_1, \ldots, z_\ell$, where $\ell \leq |x|$, such that for each j either $\chi_A(z_j) = \chi_A(z_{j-1})$ or $\chi_A(z_j) = 1 - \chi_A(z_{j-1})$, and for each j we know which of these two cases holds. And so $\chi_A(z_\ell)$ can be decided in polynomial time. $\square$

The above result is for NPMV$_t$-selectivity. A similar result holds for NPMV-selectivity. However, we obtain a weaker conclusion—membership in NP.

Theorem 5.21 *If A is Turing self-reducible and* NPMV*-selective then* $A \in$ NP.

Proof Let A be Turing self-reducible via machine M, which without loss of generality let M be such that it has some polynomial time bound that it obeys for all oracles. Via Theorem 1.24, let f be a symmetric NPMV-selector function for A. We will next give a nondeterministic recursive algorithm that, using f, accepts an input x if and only if $x \in A$.

1. In this proof, by an answer-path of deterministic polynomial-time machine M we will mean a sequence of answers (possibly correct and possibly incorrect) to its queries. Recall that x is our input. Nondeterministically guess an answer-path ρ of deterministic machine M on input x. If $M(x)$ on answer-path ρ rejects, or if after that sequence of answers the machine is not yet ready to halt in either an accepting or rejecting state but rather wishes to go on and make another query, then reject (on our own current nondeterministic path).

2. Otherwise, on our current nondeterministic path, we divide the queries implicit in ρ into two sets, YQ and NQ, for queries answered (according to ρ's claim) yes and queries answered (according to ρ's claim) no, respectively.

3. We now on our current nondeterministic path do the following. Nondeterministically guess proofs that for each ordered pair (x_1, x_2) in $(YQ \cup \{x\}) \times NQ$ it holds that $x_1 \in$ set-$f(x_1, x_2)$. Having made these guesses, if our particular nondeterministic guessing failed to yield good proofs for all those pairs (on our current path), reject (on our current path).

4. We now on our current nondeterministic path do the following. If $YQ = \emptyset$ then accept. Note that we checked that $x \in$ set-$f(x, y)$ for all $y \in NQ$. So the answer-path ρ is either correct or, if the answer for one of the strings $y \in NQ$ is incorrect, i.e., $y \in A$, then $x \in A$ follows. In both cases $x \in A$ and we may accept. If $YQ \neq \emptyset$ the algorithm next tries to prove that $YQ \subseteq A$ by isolating one of the queries in YQ and applying recursion.

5. We now on our current nondeterministic path do the following. Very loosely put, we will attempt to order the set YQ with the help of f, as was done for a P-selector in Lemma 4.7, except now we are in the more difficult case of having to work with merely an NPMV-selector, which is working on queries ρ guessed (possibly correctly, possibly incorrectly) to be in A. We'll do our ordering using nondeterminism, and will handle the elements of YQ one at a time (in any natural order).

   Initially, we take a first element from YQ and call it a_1. So, we have ordered one element.

   Continuing on, suppose we have so far (on the now-current nondeterministic path) ordered k of them so far (using perhaps quite a bit of nondeterminism to do so), and that order put things in the order $a_1, \ldots, a_k$. Take the next element of YQ; let us for now call it y. Add y to the ordered sequence $a_1, \ldots, a_k$ by guessing each of the $k + 1$ possible places one might try putting y (basically guessing a value of j between 1 and

$k+1$, and the interpretation of the value j is that we guess that y should now be placed just before the current a_j; a guess of the special value $j = k+1$ will mean we guess that y should be put right after the current a_k), and then try to prove to ourselves that it belongs in the guessed place, via guessing proofs that both (except regarding the guesses $j = 1$ and $j = k+1$ see the special comments below): (i) $a_j \in$ set-$f(y, a_j)$, as can be certified by nondeterministically guessing an appropriate path of $f(y, a_j)$ (we mention that if $a_j \in$ set-$f(y, a_j)$ holds, then it must hold that $y \in A \implies a_j \in A$), and (ii) $y \in$ set-$f(y, a_{j-1})$, as can be certified by nondeterministically guessing an appropriate path of $f(y, a_{j-1})$. Except, at the two extremes, we need only guess an appropriate one of these two paths. That is, if we guess $j = k+1$ and guess a path showing that $y \in$ set-$f(y, a_k)$, we skip making a guess regarding (i). And if we guess $j = 1$ and guess a path showing that $a_1 \in$ set-$f(y, a_1)$, we skip making a guess regarding (ii). Having made these (one or two) path guesses, if either of them that we made fails to yield the desired certificate, then the path on which we made these guesses rejects. Otherwise, we put y in the place indicated, e.g., just before the current a_j (or just after a_k if $j = k+1$), and we go on to similarly handle the next element.

On a given path that succeed in ordering all the elements of YQ in the above fashion, let us (on that path) let $a_{min}(x)$ denote the minimum a_i in its ordering. (Note then that for all i it holds that $a_{i+1} \in$ set-$f(a_i, a_{i+1})$. So $a_{min}(x) \in A$ implies $Y \subseteq A$.)

6. We now, on our current nondeterministic path, go forward to repeat this whole simulation for $a_{min}(x)$. (Crucially, since $a_{min}(x)$ was generated in the self-reduction tree, which is only polynomially deep at most, this simulation is making progress.)

For a proof that this is a correct nondeterministic polynomial-time algorithm that decides $x \in A$, we observe the following: If $x \in A$ then a correct path of M can be chosen such that $YQ \subseteq A$ and $NQ \subseteq \overline{A}$. Then certainly $y_{min}(x) \in A$ and the argument holds recursively for this $y_{min}(x)$. Moreover, for each z in the sequence $z_1, z_2, \ldots$, where $z_1 = y_{min}(x)$ and $z_{i+1} = y_{min}(z_i)$, it holds that z_{i+1} is in the self-reduction tree of z_i so there can only be polynomially many i for which the set YQ is nonempty. A correct guess of the path will return accept for the last element of this sequence and so for all elements (including x).

Next we show that there can be no accepting computation if $x \notin A$. If a correct path is chosen then this path will reject, so the algorithm will reject in step 1. If the chosen path is incorrect, then $NQ \subseteq \overline{A}$ since $x \in$ set-$f(x, y)$ for each $y \in NQ$ and $x \notin A$. So an incorrect path must mean $YQ \not\subseteq A$. However, then $y_{min}(x) \notin A$. So the recursion identifies a sequence $z_1, z_2, \ldots$ where $z_1 = y_{min}(x)$, and $(\forall i \geq 1)[z_{i+1} = y_{min}(z_i) \wedge |z_{i+1}| < |z_i| \wedge z_i \notin A]$. Since all queries on the path guessed in step 1 must have length less than the length of the input, it must hold that for at least one of these strings z_i the algorithm cannot find an accepting path. □

5.5 Reduction and Equivalence Classes

As we saw in Chapter 4, P = NP follows if all NP sets $\leq_{btt}^{p}$-reduce, or even $O(n^{1-\epsilon})$-truth-table reduce, to a P-selective set. It is a major open issue whether P = NP follows from the assumption that all NP sets $\leq_{tt}^{p}$-reduce to a P-selective set. Note, however, that if one could prove that $\{L \mid (\exists A \in \text{P-sel})[L \leq_{tt}^{p} A]\} = \{L \mid (\exists A \in \text{P-sel})[L \leq_{O(n^{1-\epsilon})-tt}^{p} A]\}$, then one would indirectly have established that P = NP follows if all NP sets reduce to a P-selective set.

In this section we study whether such collapses occur in the reducibility (and equivalence) closures of the P-selective sets. We will see that some collapses do occur, but also that many separations can be obtained—including one showing that the indirect attack sketched above is hopeless.

Our study in this section more generally is an attempt to understand the interaction between P-selective sets and the power of polynomial-time reductions: When do more powerful reductions to P-selective sets yield additional languages? When does additional reduction power yield, due to the flexibilities inherent in the class P-sel, no additional languages?

The following two notions will capture the classes that we will focus on in this section.

$$R_r(\text{P-sel}) = \{A \mid (\exists B \in \text{P-sel})[A \leq_r B]\}$$

and

$$E_r(\text{P-sel}) = \{A \mid (\exists B \in \text{P-sel})[A \equiv_r B]\},$$

for various polynomial-time reductions $\leq_r$.

5.5.1 Equalities

The following theorem is an immediate consequence of Theorem 5.12.

Theorem 5.22 $\text{P} \subsetneq \text{P-sel} = R_m^p(\text{P-sel}) = R_{ptt}^p(\text{P-sel}) = R_{pos}^p(\text{P-sel}) = E_m^p(\text{P-sel}) = E_{ptt}^p(\text{P-sel}) = E_{pos}^p(\text{P-sel}).$

A Turing reduction that is limited to k queries can query at most $2^k - 1$ different strings. On the other hand the queries $y_1, y_2, \ldots, y_{2^k-1}$ generated by a $2^k - 1$ bounded-truth-table reduction can also be listed by a Turing reduction (note that k is a constant). The Turing reduction machine can also order these queries according to the P-selector f as shown by Lemma 4.7, so that we may, without loss of generality, assume that $y_1 \leq_f y_2 \leq_f \cdots \leq_f y_{2^k-1}$. In this ordered set, the Turing machine can via binary search find the minimum i such that $y_i \in A$, using at most k queries to A. The rest of the membership queries are implicitly answered by the ordering, so that we have the following.

Proposition 5.23 *For any $k \geq 1$, $R^p_{k\text{-}T}(\text{P-sel}) = R^p_{(2^k-1)\text{-}tt}(\text{P-sel})$.*

The observation above is not limited to a constant number of queries. When the polynomial-time Turing reduction can brute-force list the truth-table queries, the sorting and querying of a limited number of queries can be performed. Thus, the following also holds.

Proposition 5.24 $R^p_{O(\log n)\text{-}T}(\text{P-sel}) = R^p_{tt}(\text{P-sel})$.

If we limit the number of questions in the reduction to a single question we have the following result.

Theorem 5.25 $R^p_{1\text{-}T}(\text{P-sel}) = R^p_{1\text{-}tt}(\text{P-sel}) = E^p_{1\text{-}T}(\text{P-sel}) = E^p_{1\text{-}tt}(\text{P-sel})$.

Proof We show that $R^p_{1\text{-}tt}(\text{P-sel}) \subseteq E^p_{1\text{-}tt}(\text{P-sel})$. Let $A \leq^p_{1\text{-}tt} B$ for some P-selective set B, and let M be a machine witnessing this reduction. Without loss of generality, we may assume that M generates a query $q(x)$ for every $x \in \Sigma^*$. The question of whether M accepts x may or may not depend on the membership of $q(x)$ in B. (By "depends on," we mean that $M(x)$'s program behavior is such that $x \in L(M^{\Sigma^*}) \iff x \notin L(M^\emptyset)$.) Let $Y = \{x \mid q(x) \in B$ and $M(x)$ depends on $q(x)\}$. Y is easily seen to be P-selective, and it is also easy to see that $Y \leq^p_{1\text{-}tt} A$ and $A \leq^p_{1\text{-}tt} Y$. $\qquad \square$

5.5.2 Inequalities

Essentially all other relationships between the reduction and equivalence classes are known to be inequalities. We first prove that $R^p_{1\text{-}tt}(\text{P-sel}) \neq \text{P-sel}$.

Theorem 5.26 $\text{P-sel} \subsetneq R^p_{1\text{-}tt}(\text{P-sel})$.

Proof We will, in our construction, specify an infinite sequence of bits, r_1, r_2, r_3, etc. The real number r will by definition be the value of the binary fraction $0.r_1 r_2 r_3 \cdots$. The set $left(r)$, i.e., the standard left cut based on r, will be important in our construction.

In our construction, we will of course specify a value for each r_j. r_{2i-1} and r_{2i} will be defined in such a way as to ensure that the ith P-selector f_i cannot be a P-selector for the set D defined as follows. Note that D itself will depend on r. The set D consist of pairs $\langle b, w \rangle$ where $b \in \{0, 1\}$ and $w \in \Sigma^*$ are such that $\langle b, w \rangle \in D$ if and only if $b = \chi_{left(r)}(w)$. Clearly, $D \leq^p_{1\text{-}tt} left(r)$ for any r we may define. Suppose, so far $r_1, \ldots, r_{2i-2}$ have been defined in our construction (initially, no r_j's have been defined, so our initial case is $i = 1$).

Let $w_0 = \langle 0, r_1 \cdots r_{2i-2}1 \rangle$ and $w_1 = \langle 1, r_1 \cdots r_{2i-2}1 \rangle$. (For the case $i = 1$, let this mean that $w_0 = \langle 0, 1 \rangle$ and $w_1 = \langle 1, 1 \rangle$.) By definition, exactly one of $\{w_0, w_1\}$ will belong to D (though which one will belong to D will depend on D, which itself depends on the r that this construction is defining). If $f_i(w_0, w_1) = w_0$ we let $r_{2i-1} = r_{2i} = 1$, and if $f_i(w_0, w_1) = w_1$ we let $r_{2i-1} = r_{2i} = 0$. It is not hard to see that this will ensure that, regardless of what happens later in the construction, f_i cannot be a P-selector for D. ❑

Note that the particular time bound on the P-selector in the proof of Theorem 5.26 is nonessential. A variant of the proof based on a listing of all (total) recursive functions rather than on a listing of the polynomial-time P-selector functions thus in fact shows that $R^p_{1\text{-}tt}(\text{P-sel})$ is not even captured by the semi-recursive sets. It is even true (and this is a slightly stronger claim, due to the fact that there exists a set X such that all total recursive functions are in FP^X) that *for all* sets X it holds that $R^p_{1\text{-}tt}(\text{P-sel}) \not\subseteq P^X$-sel.

The following theorem separates reduction classes from equivalence classes to a quite strong extent.

Theorem 5.27 $R^p_{2\text{-}tt}(\text{P-sel}) \not\subseteq E^p_T(\text{P-sel})$.

Proof We will construct a set A that is $\leq^p_{2\text{-}tt}$-reducible to a P-selective set, but is not Turing equivalent to any P-selective set. To achieve the first goal we ensure that A has only one string at distinguished lengths. A is constructed by stages where at stage s a string x_{m_s} of length m_s is added to A. A has no other strings than these strings x_{m_s} for $s \in \mathbb{N}$. The length m_s is computed by some appropriate superexponential function of s. Then A 2-truth-table reduces to the gappy left cut that has all the strings of length m_s that are lexicographically less than or equal to x_{m_s}. On input x, query x and $x + 1$. The answers to these queries will be, respectively, yes and no if and only if $x = x_{m_s}$.

To achieve the second goal, namely that A is not Turing equivalent to any P-selective set, we at stage $s = \langle i, j, k \rangle$ diagonalize against triples consisting of oracle machines M_i and M_j and selector function f_k. Since we need the selector function to induce an ordering on the query sets of these oracle machines below, we further limit ourselves to an enumeration of *symmetric* selector functions rather than of all selector functions. Clearly such an enumeration exists and clearly also, by Theorem 1.4, every set in P-sel has one of these functions as its P-selector. We will construct A in such a way as to ensure that for no P-selective set having f_k as one of its P-selector functions is it the case that both $L(M_i^X) = A$ and $L(M_j^A) = X$.

At stage s, let A_s be the set A defined so far. We either will find some string of length m_s to add to A or will set $A_{s+1} = A_s$.

We compute the union of the sets of queries produced by M_i on all inputs of length m_s and all possible oracle sets. Let $Q = \{q_1, \ldots, q_n\}$ be this set of queries. Note that $\|Q\| = n \leq 2^{(m_s)^i + 1}$. By Lemma 4.7, we without loss of generality assume that this is ordered according to f_k, i.e., $f_k(q_\ell, q_{\ell+1}) = q_{\ell+1}$ for $1 \leq \ell < n$. Now there are several cases.

Case 1: There is a p such that $M_j^{A_s}(q_p) = 1$ and $M_j^{A_s}(q_{p+1}) = 0$.

Case 2: There is an x of length m_s such that there is a p such that $M_j^{A_s \cup \{x\}}(q_p) = 1$ and $M_j^{A_s \cup \{x\}}(q_{p+1}) = 0$.

Case 3: Neither Case 1 nor Case 2, which means that the language accepted by $M_j^{A_s}$ is consistent with the ordering of the queries of M_i on inputs of length m_s (we are not in Case 1) and there is no x of length m_s such that the language of $M_j^{A_s \cup \{x\}}$ is not consistent with this ordering (we are not in Case 2).

If we are in Case 1 we do nothing and if we are in Case 2 we choose the least x satisfying Case 2 and set $A_{s+1} = A_s \cup \{x\}$. In both cases M_j cannot be a reduction from X to A for any X that has f_k as a selector.

If we are in Case 3, then for each length m_s there must be an index I such that $q_p \notin L(M_j^{A_s})$ if and only if $p \le I$ (we are not in Case 1) and for all x of length m_s there must be an index $0 \le i(x) \le n$ such that $q_p \notin L(M_j^{A_s \cup \{x\}})$ if and only if $p \le i(x)$ (we are not in Case 2).

Every string x for which $i(x) \ne I$ is queried either in the computation $M_j^{A_s}(q_I)$ or in the computation $M_j^{A_s}(q_{I+1})$. Otherwise $M_j^{A_s \cup \{x\}}(q_I) = 0$ and $M_j^{A_s \cup \{x\}}(q_{I+1}) = 1$ and consequently $i(x) = I$. As these computations can query only polynomially many strings and there are 2^{m_s} distinct strings of length m_s it follows that there are two distinct strings x and y for which $i(x) = i(y) = I$. Now if $M_i^{L(M_j^{A_s \cup \{x\}})}(x)$ accepts then $M_i^{L(M_j^{A_s \cup \{y\}})}(x)$ also accepts and we can diagonalize by letting $A_{s+1} = A_s \cup \{y\}$. If $M_i^{L(M_j^{A_s \cup \{x\}})}(x)$ rejects then we let $A_{s+1} = A_s \cup \{x\}$. $\square$

Since for any k it holds that $\mathrm{E}_{k\text{-}tt}^p(\text{P-sel}) \subseteq \mathrm{R}_{k\text{-}tt}^p(\text{P-sel})$, we obtain the following corollary.

Corollary 5.28 *For every $k \ge 2$, it holds that $\mathrm{E}_{k\text{-}tt}^p(\text{P-sel}) \subsetneq \mathrm{R}_{k\text{-}tt}^p(\text{P-sel})$ and $\mathrm{E}_{k\text{-}T}^p(\text{P-sel}) \subsetneq \mathrm{R}_{k\text{-}T}^p(\text{P-sel})$.*

Corollary 5.29 $\mathrm{E}_{btt}^p(\text{P-sel}) \subsetneq \mathrm{R}_{btt}^p(\text{P-sel})$, $\mathrm{E}_{tt}^p(\text{P-sel}) \subsetneq \mathrm{R}_{tt}^p(\text{P-sel})$, *and* $\mathrm{E}_T^p(\text{P-sel}) \subsetneq \mathrm{R}_T^p(\text{P-sel})$.

The following theorem shows that for every $k \ge 2$ the truth-table equivalence class of k is not captured by the reduction class of $k - 1$.

Theorem 5.30 *For every $k \ge 2$, $\mathrm{E}_{k\text{-}tt}^p(\text{P-sel}) \not\subseteq \mathrm{R}_{(k-1)\text{-}tt}^p(\text{P-sel})$.*

We first need some notions from coding theory. A *k-walk* is a sequence $b_1, \ldots, b_m \in \{0,1\}^k$ such that from each bitstring to the next exactly one bit changes. The number of times the bit at the ith position changes is called the *transition count* of i. A walk is *self-avoiding* if it visits no bitstring twice.

For example $001, 011, 010, 110, 111$ is a self avoiding 3-walk with transition counts 1, 1, and 2 for $i = 1, 2, 3$, respectively. The transition count of a k-walk is the maximum of the transition counts taken over all positions $1, \ldots, k$.

Lemma 5.31 *For each k and n satisfying $kn+1 < 2^n$ there exists an n-walk $b_1, \ldots, b_m$ with transition count at most $k+1$ such that for no n-walk $b'_1, \ldots, b'_\ell$ with transition count at most k do we have $\{b_1, \ldots, b_m\} \subseteq \{b'_1, \ldots, b'_\ell\}$.*

Proof Let $b_1, \ldots, b_\ell$ be an n-walk with transition count at most k visiting a maximum number of bitstrings. Clearly $\ell \leq nk + 1 < 2^n$. There must thus exist a bitstring b that is not visited. Extend the walk from b_ℓ to b by stepwise flipping the bits where b_ℓ and b differ. This gives a walk with transition count at most $k + 1$, the cardinality of whose set of visited bitstrings is one more than that of any n-walk having transition count at most k. $\Box$

Lemma 5.32 *For each k there exists a self-avoiding k-walk whose transition counts are exactly k for all but the last position, and whose last position has a transition count of 1.*

Proof For $k = 1$ such a walk is $0, 1$. For the inductive step, let the k-walk $b_1, \ldots, b_\ell$ with the desired properties already be constructed. The $(k + 1)$-walk starts with $b_1 0, b_2 0, \ldots, b_\ell 0$. Then it goes on to $b_\ell 1$, which will ensure that the following bitstrings will not have been visited before. We now toggle position k, then position 1, then position k once more, then position 2, then position k again, then position 3, and so forth up to position $k - 1$, followed by a final flip of position k.

Clearly, the constructed $k + 1$ walk is self-avoiding. The transition counts for the first $k-1$ positions are all $k+1$. For the kth position it is $1+(k-1)+1 = k + 1$. For the last position it is 1. $\Box$

Now we prove Theorem 5.30.

Proof of Theorem 5.30   We construct by stages a tally set T and a P-selective set A such that $T \equiv^p_{k\text{-}tt} A$, but for every pair s_1, s_2, the $(k-1)$-truth-table reduction M_{s_1} and selector function f_{s_2} fail to witness that $T \leq^p_{(k-1)\text{-}tt} X$ for any X that has selector function f_{s_2}.

We describe the strings put into A_s and T_s at stage s, where $s = \langle s_1, s_2 \rangle$. Let n be some suitable length, i.e., one such that adding strings to A and T at length n cannot undo the achievements of previous stages and, moreover, such that membership in A of strings put into A at previous stages can be computed in time linear in n. For $i = 1, \ldots, k$, let $x_i = 0^{n+i}$. Let $y_1, \ldots, y_{k^2-k+1}$ be the lexicographically first $k^2 - k + 1$ strings of length n and let $c_1, \ldots, c_{k^2-k+2}$ be the k-walk constructed in Lemma 5.32

Let $Q = \{q_1, \ldots, q_m\}$ be the union of the sets of strings queried by M_{s_1} on inputs $x_1, \ldots, x_k$. We assume Q to be ordered as per Lemma 4.7 by P-selector f_{s_2} so that $f_{s_2}(q_r, q_{r+1}) = q_{r+1}$ and that $\|Q\| \leq k(k - 1)$. Now let $b_i^t = 1$ if and only if M_{s_2} accepts x_i with oracle $T_{s-1} \cup \{q_t, \ldots, q_m\}$, and consider $B^t = \langle b_1^t, \ldots, b_k^t \rangle$. Since $m \leq k^2 - k$ and $c_1, \ldots, c_{k^2-k+2}$ are all distinct, there must by at least one a such that $c_a \neq B_t$ for all t. We take the least such a and put x_r into T_s if and only if the rth bit of c_a is 1, and put $y_1, \ldots, y_a$ into A_s.

Since $c_a \neq B^t$ for any t it holds for any set X with selector f_{s_2} that there is an index r such that $x_r \in T$ if and only if M^X rejects x_r.

It remains to show that $A \equiv^p_{k\text{-}tt} T$. To find out whether $x_r \in T$, we just need to know how often a bitflip occurred at the rth position in the walk $c_1, \ldots, c_a$. We find this out by querying exactly those k strings in $\{y_1, \ldots, y_{k^2-k+1}\}$ that correspond to steps in the walk $c_1, \ldots, c_{k^2-k+2}$ where the rth position changed. For the other direction, note that from the knowledge of c_a we can easily reconstruct a, which means that we can determine a by querying $x_1, \ldots, x_k$ to T. We conclude that T and A are k-equivalent.   ☐   Theorem 5.30

Corollary 5.33 $R^p_{(k-1)\text{-}tt}(\text{P-sel}) \subsetneq R^p_{k\text{-}tt}(\text{P-sel})$, $E^p_{(k-1)\text{-}tt}(\text{P-sel}) \subsetneq E^p_{k\text{-}tt}(\text{P-sel})$, and $R^p_{(k-1)\text{-}T}(\text{P-sel}) \subsetneq R^p_{k\text{-}T}(\text{P-sel})$.

Regarding the relationship between $R^p_{tt}(\text{P-sel})$ and $E^p_T(\text{P-sel})$, we have the following theorem.

Theorem 5.34 $E^p_T(\text{P-sel}) \not\subseteq R^p_{tt}(\text{P-sel})$.

Proof We construct a set $L = \bigcup_i L_i$ that is Turing equivalent to a P-selective set $A = \bigcup_i A_i$, yet is not truth-table reducible to any P-selective set. We make use of the fact that a polynomial-time Turing reduction can perform binary search on the 2^n strings of length n. However, each truth-table reduction can only query a (fixed) polynomial number of strings and so can be diagonalized against. The set A will again be a gappy left cut. At length n, where we diagonalize against machine M_i and P-selector f_j, A will consist of all strings that are lexicographically smaller than some string w that is determined as follows.

Let $\{x_1, x_2, \ldots, x_n\} = \{0^{n+1}, 0^{n+2}, \ldots, 0^{n+n}\}$. Let $Q = \{q_1, q_2, \ldots, q_m\}$ be the union of the sets of strings queried by M_i on inputs in $\{x_1, x_2, \ldots, x_n\}$, ordered by selector function f_j according to Lemma 4.7, so that $q_r \leq_{f_j} q_{r+1}$. Let $\{w_1, w_2, \ldots, w_m\}$ be the set of strings of length n such that the kth bit of w_ℓ equals 1 if and only if M_i accepts x_k with oracle $\{q_\ell, \ldots, q_m\}$, and let w be the lexicographically least string of length n not equal to any of these w_ℓ. Such a string always exists, except for perhaps a finite number of exceptions, since m is polynomially bounded in n. Then put x_k in L_{i+1} if and only if the kth bit of w equals 1, and put all strings of length n lexicographically less than or equal to w into A_{i+1}. L cannot truth-table reduce to any P-selective set with selector f_j via M_i. On the other hand for any given length n the string w can easily be recovered from oracle A in polynomial time using binary search, so $L \leq^p_T A$. For length n the string w itself can be recovered by querying L about the strings $0^{n+1}, \ldots, 0^{2n}$, so $A \leq^p_T L$.   ☐

The proof of this theorem can be modified to diagonalize against bounded-truth-table reductions while maintaining equivalence on the truth-table scale. As a corollary to the proof of this theorem we thus note the following.

Corollary 5.35 $E_{tt}^p(\text{P-sel}) \not\subseteq R_{btt}^p(\text{P-sel})$.

The following corollary summarizes the results obtained thus far.

Corollary 5.36

1. $R_{btt}^p(\text{P-sel}) \subsetneqq R_{tt}^p(\text{P-sel}) \subsetneqq R_T^p(\text{P-sel})$.
2. $E_{btt}^p(\text{P-sel}) \subsetneqq E_{tt}^p(\text{P-sel}) \subsetneqq E_T^p(\text{P-sel})$.
3. $E_{btt}^p(\text{P-sel}) \subsetneqq R_{btt}^p(\text{P-sel})$.
4. $E_{tt}^p(\text{P-sel}) \subsetneqq R_{tt}^p(\text{P-sel})$.
5. $E_T^p(\text{P-sel}) \subsetneqq R_T^p(\text{P-sel})$.
6. $E_{tt}^p(\text{P-sel})$ *and* $R_{btt}^p(\text{P-sel})$ *are incomparable.*
7. $E_T^p(\text{P-sel})$ *and* $R_{btt}^p(\text{P-sel})$ *are incomparable.*
8. $E_T^p(\text{P-sel})$ *and* $R_{tt}^p(\text{P-sel})$ *are incomparable.*

The next theorem establishes a relationship between truth-table equivalence classes and Turing equivalence classes.

Theorem 5.37 *For each* $k \geq 1$, $E_{2k\text{-}tt}^p(\text{P-sel}) \not\subseteq E_{(k-1)\text{-}T}^p(\text{P-sel})$.

Proof Fix any $k \geq 0$. We will construct a set $T = \bigcup_n T_n$ and a set $A = \bigcup_n A_n$ in stages such that $T \equiv_{2k+2\text{-}tt}^p A$. For each n we will let $\widehat{T}_n = \bigcup_{\ell < n} T_\ell$. For any pair of k-Turing reductions M_i and M_j and symmetric P-selector f_ℓ and any set X that is P-selective by f_ℓ, the construction will enforce that either M_i fails to witness that $T \leq_{k\text{-}T}^p X$ or M_j fails to witness that $X \leq_{k\text{-}T}^p T$. A transition from k to $k-1$ will then give the exact statement of the theorem.

At stage $n = \langle i, j, \ell \rangle$ we will choose a new length $s(n)$ from some increasing sequence such that strings of length $s(n)$ cannot on inputs of length $s(n-1)$ or less be queried by any of the machines that were considered at stage $n-1$ or earlier. Moreover, $s(n)$ will be so large that for strings of length at most $s(n-1)$ membership in A can be decided in time linear in $s(n)$. The construction of A will be constructive in such a way that it will be clear not only that such a sequence exists but also that it is easily computable in the sense that for any m the question of whether $m = s(n)$ for some n can be decided in linear time.

Let $x_1, \ldots, x_{2k+2}$ be the lexicographically first $2k+2$ strings of length $s(n)$. Both A_n and T_n will consist of a subset of these strings.

It will be the case that either $A_n = T_n = \emptyset$ or there is some t, $1 \leq t \leq 2k+2$, such that $A_n = \{x_1, \ldots, x_t\}$ and $T_n = \{x_t\}$. Since this is true for all stages, it follows that $T \equiv_{2k+2\text{-}tt}^p A$. We will now turn to the diagonalization part.

Let Q be the set of all strings that can possibly by queried on inputs $x_1, \ldots, x_{2k+2}$ by machine M_i. We assume that M_i is "query-clocked" to never ask more than k oracle queries. Then $\|Q\| \leq (2k+2)(2^k-1)$, the $2^k - 1$ accounting for the implicit tree of possible queries due to later queries depending on the answers to earlier queries. If Q happens to be empty, then if $M_i^\emptyset(x_1)$ accepts we let $A_n = T_n = \emptyset$ and if it rejects we let $A_n = T_n = \{x_1\}$. In this case diagonalization is easy.

We now assume $Q = \{q_1, \ldots, q_m\}$ for $m \geq 1$. As before we assume that, for $1 \leq r \leq m-1$, $q_r \leq_{f_\ell} q_{r+1}$ (via Lemma 4.7). Now if a P-selective set X exists for which f_ℓ is a symmetric selector function and such that X satisfies $q_r \in X \cap Q$, then necessarily $\{q_r, \ldots, q_m\} \subseteq X$.

For oracles $\widehat{T}_{n-1} \cup \{x_r\}$, $1 \leq r \leq 2k+2$, we define a cut-off point c_r as the minimum integer $1 \leq c_r \leq m$ such that $M_j^{\widehat{T}_{n-1} \cup \{x_r\}}$ accepts q_{c_r}. It could be the case that there is no such string. In that case we set $c_r = m+1$ and $q_{m+1} = \top$, where $\top$ is a special string such that $q_r \leq_{f_\ell} \top$ for every $q_r \in Q$, and just say that $M_j^{\widehat{T}_{n-1} \cup \{x_r\}}$ accepts $\top$ for the flow of the proof, but of course really meaning that $M_j^{\widehat{T}_{n-1} \cup \{x_r\}}$ accepts no string in Q.

The assumption that M_j reduces to T a P-selective set that obeys f_ℓ will now imply that if $T_n = \{x_r\}$ then $M_j^{\widehat{T}_{n-1} \cup \{x_r\}}$ must accept *all* q_t for $t \geq c_r$. If M_j behaves differently for some x_r—that is, it accepts some q_t and rejects q_{t+1} with oracle $\widehat{T}_{n-1} \cup \{x_r\}$—we let $T_n = \{x_r\}$ and $A_n = \{x_1, \ldots, x_r\}$. In that case, diagonalization is again easy, so we assume this does not occur.

We now claim that there are $x_r \neq x_t$ such that $q_{c_r} = q_{c_t}$. This can be seen as follows. Assume that all $2k+2$ of the q_{c_i} are distinct and let $\pi : \{x_1, \ldots, x_{2k+2}\} \to Q \cup \{\top\}$ be an injective mapping that gives these strings. Let $P = \{p_1, \ldots, p_{2k+2}\} \subseteq Q \cup \{\top\}$ be the range of π, such that $p_s \leq_{f_\ell} p_{s+1}$ for $1 \leq s \leq 2k+1$. Let $y_s = \pi^{-1}(p_s)$.

Consider p_{k+1}. By assumption $M_j^{\widehat{T}_{n-1} \cup \{y_s\}}$ accepts p_s and so it must accept p_{k+1} if $s \leq k+1$. Also, p_s is the minimum string in the order induced by f_ℓ that is accepted by $M_j^{\widehat{T}_{n-1} \cup \{y_s\}}$. So $M_j^{\widehat{T}_{n-1} \cup \{y_s\}}$ must reject p_{k+1} if $s > k+1$.

$M_j^{\widehat{T}_{n-1}}$ either accepts or rejects p_{k+1}. If it accepts, it must query every string in $\{y_{k+2}, \ldots, y_{2k+2}\}$, since if it did not ask some y_s with $s > k+1$ then $M_j^{\widehat{T}_{n-1} \cup y_s}(p_{k+1})$ would do the same as $M_j^{\widehat{T}_{n-1}}(p_{k+1})$, namely accepting, contradicting that $M_j^{\widehat{T}_{n-1} \cup y_s}$ must reject p_{k+1}. If $M_j^{\widehat{T}_{n-1}}$ rejects, it must query every string in $\{y_1, \ldots, y_{k+1}\}$. In either case it must make at least $k+1$ queries.

Now select $x_r \neq x_s$ such that $q_{c_r} = q_{c_s}$. Then $M_j^{\widehat{T}_{n-1} \cup \{x_r\}}$ and $M_j^{\widehat{T}_{n-1} \cup \{x_s\}}$ must accept the same, possibly empty, subset of Q. Let Y be this set.

If $\widehat{T}_n$ is either $\widehat{T}_{n-1} \cup \{x_r\}$ or $\widehat{T}_{n-1} \cup \{x_s\}$ then for any x_k it is true that $x_k \in T_n$ if and only if $x_k \in L(M_j^Y)$. However, $\widehat{T}_{n-1} \cup \{x_r\} \neq \widehat{T}_{n-1} \cup \{x_s\}$, so M_j^Y cannot be correct on *both* these sets. We can now choose T_n to be either $\{x_r\}$ or $\{x_s\}$ such that M_j^Y either rejects a string in $\widehat{T}_n$ or accepts a string not in $\widehat{T}_n$. $\square$

Corollary 5.33 states that $R^p_{(k-1)\text{-}T}(\text{P-sel}) \subsetneq R^p_{k\text{-}T}(\text{P-sel})$. In the non-constant case such a hierarchy can also be established.

Lemma 5.38 *For every k, $R^p_{n^k\text{-}T}(\text{P-sel}) \subsetneq R^p_{n^{k+1}\text{-}T}(\text{P-sel})$.*

Proof We construct sets A and B by stages so that A is not reducible to any P-selective set by a machine that makes only n^k adaptive queries and so that A is reducible to B via a machine that we will define.

Let $A_0 = \emptyset$, $A = \bigcup_s A_s$, $B_0 = \emptyset$, and $B = \bigcup_s B_s$. At stage $s = \langle i, j \rangle$ of the construction we diagonalize against reduction M_i and P-selector f_j. First we pick a suitable length $\ell(s)$ such that adding strings of length $\ell(s)$ to either A or B does not interfere with any actions taken earlier and such that $\ell(s)^{k+1}2^{\ell(s)^k} + 1 < 2^{\ell(s)^{k+1}}$. Now let $x_1, x_2, \ldots, x_{\ell(s)^{k+1}}$ be the lexicographically first $\ell(s)^{k+1}$ strings of length $\ell(s)$. On inputs $x_1, x_2, \ldots, x_{\ell(s)^{k+1}}$, machine M_i can generate at most $\ell(s)^{k+1}2^{\ell(s)^k}$ different queries in the union Q of all query sets generated on all these inputs taken over all possible oracle sets. Sort these queries according to the P-selector (Lemma 4.7). Let X be a P-selective set with P-selector f_j, and let w_X be the characteristic string of $x_1, \ldots, x_{\ell(s)^{k+1}}$ in $L(M_i^X)$. Since there are at most $\|Q\| + 1 \leq \ell(s)^{k+1}2^{\ell(s)^k} + 1$ possibilities for $Q \cap X$, it follows that there are also at most $\ell(s)^{k+1}2^{\ell(s)^k} + 1$ possibilities for w_X when X varies over all P-selective sets that obey f_j. However, there are $2^{\ell(s)^{k+1}}$ different possible settings of $x_1, x_2, \ldots, x_{\ell(s)^{k+1}}$ and so we can choose the first of these settings that does not agree with f_j, and let that setting set the membership of $x_1, x_2, \ldots, x_{\ell(s)^{k+1}}$ in A, so that M_i fails to witness an $\ell(s)^k$ reduction to any P-selective set with selector f_j. On the other hand characteristic sequence of this setting in A is a number less than or equal to $2^{\ell(s)^{k+1}}$, call it d. If we let B_s consist of the lexicographically first d strings of length $\ell(s)$, then a binary searching oracle machine with oracle B can recover this setting. □

Summing up, we get the following.

Corollary 5.39 *The following strict containments hold:*

$$R^p_{tt}(\text{P-sel}) \subsetneq R^p_{n\text{-}T}(\text{P-sel}) \subsetneq R^p_{n^2\text{-}T}(\text{P-sel}) \subsetneq \cdots \subsetneq R^p_T(\text{P-sel}).$$

5.6 Bibliographic Notes

Hemaspaandra and Jiang [HJ95] proved the boolean (non)closures of P-sel under various connectives. The results of that paper appear as the theorems and corollaries of Section 5.2, although the paper does not mention the (straightforward) implications of the proofs for other types of selectivity than P-selectivity. The present strong form of Theorem 5.5 is due to Tantau [Tan01].

The notion of positive reductions, known from recursion theory [Joc79], was introduced to complexity theory by Selman [Sel82a]. Selman also proved Lemma 5.14 [Sel82a]. Buhrman, Torenvliet, and van Emde Boas [BTvEB93] extended this closure of P-sel to Theorem 5.12, from which result the equality parts of Theorem 5.22 follow (Theorem 5.22's inequality follows from, for example, Theorem 1.16). Theorem 5.12 is about positive reductions, but is in fact about *globally* positive reductions. That is, it must be a property of the machine that it computes a positive reduction relative to *all* oracles. For sets A and B to have $A \leq^p_{\text{"positive"}} B$, one could conceivably just demand that adding strings to B can only lead to acceptance of more strings. Such reductions (and their analogs regarding removing strings) are called *locally* positive. Locally positive reductions were introduced and studied by Hemachandra and Jain [HJ91]. For locally positive reductions the issue of closure under positive Turing reductions is still open.

The extension of Theorem 5.12 to NPSV_t-selectivity (Lemma 5.16 and Theorem 5.17) is due to Hemaspaandra et al. [HHN+95].

Buhrman and Torenvliet [BT96] proved the closure of P-selectivity under the combination of self-reducibility and a 1-truth-table reduction; this appears here as Theorem 5.18. Theorem 5.20, which extends the behavior to nondeterministic classes and reductions, is due to Hemaspaandra et al. [HHN+95]. Theorem 5.21 is due to Hemaspaandra et al. [HNOS96a]. Beigel, Kummer, and Stephan [BKS95a] provide a relativized world where the closure of Theorem 5.18 seems to be about as far as one can go, i.e., a world in which there exists a disjunctively self-reducible set in NP − P that is $\leq^p_{2\text{-}tt}$-reducible to a P-selective set in NP.

Hemaspaandra, Hoene, and Ogihara [HHO96] broadly studied equalities and inequalities of reduction and equivalence classes based on P-selective sets, and Proposition 5.23, Theorems 5.25, 5.26, and 5.27, and Corollaries 5.28 and 5.29 are due to that paper.

Theorem 5.30 is also due to Hemaspaandra, Hoene, and Ogihara [HHO96]. We give here an alternate proof (involving also two lemmas, Lemmas 5.31 and 5.32) due to Tantau ([Tan01], see also [Tan00]). Tantau [Tan00] proves a stronger form of Theorem 5.30, namely, he shows that the relationship $E^p_{k\text{-}tt}(\text{P-sel}) \not\subseteq R^p_{(k-1)\text{-}tt}(\text{P-sel})$ still holds if we relativize only the right-hand side. That is, for all oracles X we have that $E^p_{k\text{-}tt}(\text{P-sel}) \not\subseteq R^{p,X}_{(k-1)\text{-}tt}(\text{P-sel})$. Furthermore, it is also shown there that $E^p_{k\text{-}T}(\text{P-sel}) \not\subseteq R^p_{(2^k-2)\text{-}tt}(\text{P-sel})$ (i.e., that the result $E^p_{k\text{-}T}(\text{P-sel}) \subseteq R^p_{(2^k-1)\text{-}tt}(\text{P-sel})$ [HHO96], which follows for example from Proposition 5.23, is tight).

Corollary 5.33, Theorem 5.34, and Corollaries 5.35 and 5.36 are also due to Hemaspaandra, Hoene, and Ogihara [HHO96].

Theorem 5.37 was originally stated in a stronger form by Hemaspaandra, Hoene, and Ogihara, but that version was retracted and replaced by the version included here (see the erratum to [HHO96]). It is an open question whether the original claim can be proven. We conjecture that it can.

6. Generalizations and Related Notions

In this chapter we will discuss some generalizations and refinements of P-selectivity along with some other notions related to P-selectivity. Since these notions are not this book's central focus, we will not give proofs here, but rather will present definitions and explain the concepts and their relationships to the forms of selectivity discussed in previous chapters. We will also mention some—but certainly not all—results about these related notions.

Section 6.1 discusses such generalizations of P-selectivity as weak selectivity, multiselectivity, membership comparability, approximability, and selectivity via probabilistic selector functions. Section 6.2 discusses the P-semi-rankable sets. This class is provably a refinement of the P-selective sets. However, we will note that it remains an open research question whether all P-selective sets can be approximated by P-semi-rankable sets. Section 6.3 concludes the book by returning to an aspect of P-selectivity that was first raised in Section 1.1: the algebraic properties of the P-selective sets. In particular, Section 6.3 discusses the class of all associatively P-selective sets. This class has the property that P-sel $\subseteq$ P/linear unless the class is a strict refinement of the P-selective sets.

6.1 Generalizations of Selectivity

6.1.1 Weak Selectivity

A class of sets known as the weakly P-selective sets played an important role in the early development of selectivity theory. This class generalizes the P-selective sets by being more flexible in terms of the ordering and equivalence relations on which the "selector" function operates. A hint of the flavor of the weakly P-selective sets is given by considering sets that are a disjoint union of P-selective sets. Though such sets in general need not be P-selective, they do possess many of the properties commonly associated with the P-selective sets. For example, such sets clearly belong to P/poly, and no such set can be $\leq_m^p$-hard for NP (respectively PSPACE) unless P $=$ NP (respectively, P $=$ PSPACE). Similarly, such sets have the property that all disjunctively self-reducible sets that $\leq_m^p$-reduce to them belong to P.

6.1.2 Multiselectivity: The S(k) Hierarchy

Multiselectivity is a generalization of selectivity to sequences of strings instead of pairs of strings. The standard notion of selectivity of a set is characterized by selector functions that output *one* string in the set if at least one of the two input strings is in the set. A set A is called (i,j)-selective if there exists a polynomial-time computable function that outputs j strings in A if at least i of the arguments of the function are in A. As it turns out, the only parameter of importance is the *difference* between i and j. Thus, though more general notions may be defined, the most interesting case can be limited to the definition of the S(k) hierarchy, where S(k) is defined as follows.

Definition 6.1   *For each $k > 0$, S(k) is the class of all sets L for which there exists a polynomial-time computable function f such that, for each $n > 1$ and any distinct input strings $y_1, \ldots, y_n$,*

1. *$f(\langle y_1, \ldots, y_n \rangle) \in \{y_1, \ldots, y_n\}$ and*
2. *$\|L \cap \{y_1, \ldots, y_n\}\| \geq k \implies f(\langle y_1, \ldots, y_n \rangle) \in L$.*

It is easy to see that S(1) = P-sel.

One can vary each S(k) to form a related class that we will call fair-S(k). The fair-S(k) classes force a certain behavior on the selector function only under a promise. In particular, only if all arguments to the selector function have the same length is it required that the output of the selector function fulfills the requirements given above. For nonconstant thresholds (i.e., a threshold on the number of arguments *in* the set that forces the function to deliver a result *in* the set) we use n rather than k. This way we get, e.g., S($\log n$), S($\sqrt{n}$), S($n - 1$), etc. We denote $\bigcup_{k \geq 1}$ S(k) by SH. This hierarchy turns out to be proper.

Theorem 6.2

1. *For each $k \geq 1$, S(k) $\subsetneq$ S($k + 1$).*
2. *SH $\subsetneq$ fair-S($n - 1$).*

Though they are a direct generalization of the P-selective sets, the sets in this "multiselectivity" hierarchy have some rather distinctive properties. (Note in particular that the analogs of each part of the following theorem fail for S(1), i.e., for P-sel.)

Theorem 6.3

1. *For each $k \geq 2$, S(k) $\subsetneq$ R$_m^p$(S(k)).*
2. *There exists a set A in S(2) such that $A \leq_m^p \overline{A}$ yet $A \notin$ P.*

A set A is *autoreducible* if for some deterministic polynomial-time Turing machine M we have that (a) $A = L(M^A)$ and (b) for no input x does

$M^A(x)$ query x. From part 2 of Theorem 6.3, it follows that $S(2)$ contains an autoreducible set that is not in P. In contrast, any set that is in SH and that is reducible to itself via a *length-increasing* reduction (i.e., a reduction that outputs only strings that are longer than the input, which is of course always an autoreduction) is P-selective.

Each set in the multiselectivity hierarchy has small circuits (i.e., SH $\subseteq$ P/poly) and so as a corollary is in $\mathrm{EL}_{\Theta_2^p}$. Moreover, any set in NP $\cap$ fair-$S(n-1)$ is in the lowness class $\mathrm{L}_{\Sigma_2^p}$. However, NP $\subseteq$ fair-$S(n-1)$ if and only if P = NP.

In Chapter 3, we discussed the placement of the P-selective sets in the extended low hierarchy. The following theorem gives the optimal, in light of SH $\subseteq$ $\mathrm{EL}_{\Theta_3^p}$, relationship between the extended low hierarchy and the multiselectivity hierarchy.

Theorem 6.4

1. SPARSE $\cap$ $S(2)$ $\not\subseteq$ $\mathrm{EL}_{\Sigma_2^p}$.
2. co-SPARSE $\cap$ $S(2)$ $\not\subseteq$ $\mathrm{EL}_{\Sigma_2^p}$.

We mention in passing a result that was originally obtained hand in hand with the study of multiselectivity. This nonintuitive complexity result demonstrates the difference between the complexity measures "placement in the extended low hierarchy" and "complexity via reductions."

Theorem 6.5 $(\exists A, B)[A \notin \mathrm{EL}_{\Sigma_2^p} \wedge B \notin \mathrm{EL}_{\Sigma_2^p} \wedge A \oplus B \in \mathrm{EL}_{\Sigma_2^p}]$.

Theorem 6.5 says that the disjoint union can actually lower complexity, at least in the sense of lowering the placement within the extended low hierarchy. In contrast, note that every level of the low hierarchy is clearly closed under $\leq_m^p$ reductions. Thus, the low-hierarchy analog of Theorem 6.5 is false; in fact, it holds that: $(\forall k > 0)(\forall A, B)[A \notin \mathrm{L}_{\Sigma_k^p} \vee B \notin \mathrm{L}_{\Sigma_k^p} \implies A \oplus B \notin \mathrm{L}_{\Sigma_k^p}]$.

$\mathrm{EL}_{\Sigma_2^p}$ has other surprising behaviors. Most particularly, it is known to lack a variety of boolean closure properties; it is not closed under intersection, union, exclusive-or, or nxor.

6.1.3 Membership Comparable Sets

A selector function for a P-selective set A is a function that, for any pair of strings x_1 and x_2, rules out one of the four possibilities for $\langle \chi_A(x_1), \chi_A(x_2) \rangle$. One generalization of that concept would be a polynomial-time computable function that rules out one of the 2^m possibilities for $\langle \chi_A(x_1), \ldots, \chi_A(x_m) \rangle$. In the most general setting the number m need not be a constant, but rather may be a function of the maximum of the lengths of $x_1, \ldots, x_m$.

Definition 6.6   1. *A function f is called a g-membership comparing function (a g-mc-function, for short) for A if, for every $x_1, \ldots, x_m$ with $m \geq g(\max\{|x_1|, \ldots, |x_m|\})$, it holds that $f(x_1, \ldots, x_m) \in \{0,1\}^m$ and*

$f(x_1, \ldots, x_m) \neq \chi_A(x_1)\chi_A(x_2) \cdots \chi_A(x_m)$, *where this last expression denotes the concatenation of those m values.*

2. *A set A is* polynomial-time g-membership comparable *if there exists a polynomial-time g-mc-function for A.*

3. P-mc(g) *denotes the class of all polynomial-time g-membership comparable sets. For $\bigcup_{g \in \{1,2,3,\ldots\}}$ P-mc(g) we write P-mc(const). (Here, we are using 1 as a shorthand for the function $\lambda n.1$, and are also using the same type of shorthand for 2, 3, etc.)*

Many results known for P-selective sets also hold for membership comparable sets. For example, Theorem 2.29 applies to the membership comparable sets in the following way, due to the fact that P-mc(const) $\subseteq$ P/poly.

Theorem 6.7 *If* $\mathrm{SAT} \in \mathrm{P}^{\mathrm{P\text{-}mc(const)}}$ *then* $\mathrm{PH} \subseteq \mathrm{ZPP}^{\mathrm{NP}}$.

For stronger reductions, we can bring the collapse even further down, as is also the case with P-selective sets. For example, it holds that if $\mathrm{SAT} \in \mathrm{R}^p_{btt}(\mathrm{P\text{-}mc(const)})$ then $\mathrm{P} = \mathrm{NP}$. In fact, the currently strongest results extend this even to some truth-table reductions that allow more than a constant number of queries.

Theorem 6.8 *If* $\mathrm{SAT} \leq^p_{n^\alpha\text{-}tt}$*-reduces to some* P-mc($k$) *set for some* $\alpha < \frac{1}{k-1}$, *where $k > 1$, then* $\mathrm{P} = \mathrm{NP}$.

A generalization of Theorem 6.8 to $\leq^p_{tt}$ reductions—or even to $\leq^p_{n\text{-}tt}$ reductions—is, at the time of this writing, not in sight, nor does the proof technique seem likely to easily yield such a result. On the other hand, there is no known oracle world in which $\mathrm{P} \neq \mathrm{NP}$ and $\mathrm{SAT} \in \mathrm{R}^p_{tt}(\mathrm{P\text{-}sel})$ (or even in which $\mathrm{P} \neq \mathrm{NP}$ and SAT is $\mathcal{O}(\log n)$ membership comparable).

The following notions are related to membership comparability, in most cases being refinements of that notion.

Definition 6.9

1. *Near-Testable Sets: A set A is* near-testable *if there is a polynomial-time function that, on each input x, determines the value of $\chi_A(x) \oplus \chi_A(x+1)$, where $x + 1$ denotes the string immediately following x in lexicographical order.*

2. *Easily Countable Sets: For each $k \geq 1$, we say that a set A is* easily k-countable *if there exists a k-argument polynomial-time function f such that, for each $x_1, \ldots, x_k \in \Sigma^*$, it holds that $f(x_1, \ldots, x_k) \in \{0, 1, \ldots, k\}$ and $f(x_1, \ldots, x_k) \neq \|A \cap \{x_1, \ldots, x_k\}\|$.*

3. *$(a, b)_p$ Verbose Sets: A set A is called $(a, b)_p$ verbose if we can compute in polynomial time for each b-tuple $(x_1, \ldots, x_b)$ a set $D \subseteq \{0, 1\}^b$, with $\|D\| = a$, that contains $\chi_A(x_1)\chi_A(x_2) \cdots \chi_A(x_b)$ (the string formed by the sequence of values of the characteristic functions).*

4. $(a,b)_p$ *Recursive Sets: A set is $(a,b)_p$-recursive if there is a polynomial-time computable function f such that for each $x_1, \ldots, x_b \in \Sigma^*$ it holds that the bitstrings $f(x_1, \ldots, x_b) \in \{0,1\}^b$ and $\chi_A(x_1)\chi_A(x_2)\cdots\chi_A(x_b)$ agree in at least a positions.*

5. P-*Superterseness: A set is P-superterse if it is not $(2^{b-1}, b)_p$-verbose.*

6. *Cheatable Sets: A set is k-cheatable if it is $(k,k)_p$-verbose. A set is cheatable if it is k-cheatable for some k.*

These notions are of course closely related both to P-selectivity and to each other. As an example, the following holds.

Theorem 6.10 *Let $A \subseteq \Sigma^*$ and let $k > 0$.*

1. *If A is k-cheatable then A is easily 2^k-countable.*
2. *If A is easily k-countable then A is $(2k-1)$-membership comparable (i.e., is $(\lambda n.(2k-1))$-membership comparable).*

Near-testability is a bit different in flavor than the other notions in Definition 6.9, which are all generalizations of P-selectivity. Since near-testable sets are all in $E \cap PSPACE$, there are P-selective sets that are not near-testable. On the other hand, if even one near-testable set is not P-selective then clearly $P \neq PSPACE$, since all P sets are P-selective.

6.1.4 Probabilistic Selector Functions

Instead of choosing a polynomial-time computable selector function, we may choose the selector function from a different, more powerful class of functions. Definition 1.20 makes clear how to interpret such selectors, and we have, in the various notions of nondeterministic selectivity, already seen examples of this. A historically early example of this is of course the case where the function is recursive, resulting in the semi-recursive sets. One may also consider FEXP-selective sets and classes defined by all other sorts of selector functions. An interesting definition exists for probabilistic selectivity.

Definition 6.11 *Let an FPP-function be a function f for which a balanced (i.e., one that for some polynomial p has exactly $2^{p(|x|)}$ computation paths on each input x) nondeterministic Turing machine exists that on each input x has $f(x)$ as the output of more than half of its computation paths. A set A is FPP-selective if it has a selector function belonging to FPP.*

The following theorem holds.

Theorem 6.12 *Each set in PP is FPP-selective.*

Even sets having such powerful selector functions can be recognized in probabilistic polynomial time with the help of a modest amount of advice.

Theorem 6.13 *All FPP-selective sets are in PP/poly.*

6.2 P-Semi-Rankability: A Provable Refinement of P-Selectivity

The next notion that we will discuss in this chapter is not a generalization of P-selectivity but rather is a refinement of P-selectivity. First define $\mathrm{rank}_A(x)$, the *rank* of a string x with respect to a set A, as the number of $y \in A$ such that $y \leq_{\mathrm{lex}} x$ (i.e., the number of strings in A that precede or tie x in lexicographical order). A set A can then be defined to be *weakly P-rankable* (notated $A \in$ weakly-P-rankable) if there is a polynomial-time function f that, on each $x \in A$, computes $\mathrm{rank}_A(x)$. Note that, on inputs $x \notin A$, such a ranking function f can output any lie (or truth) that it chooses to output.

Definition 6.14

1. *A set A is* polynomial-time semi-rankable *if and only if there is a polynomial-time computable function f such that, for every x and y,*
   a) $(\exists n)[f(x,y) = \langle x, n \rangle \vee f(x,y) = \langle y, n \rangle]$ *and*
   b) $\{x,y\} \cap A \neq \emptyset \implies [(x \in A \wedge f(x,y) = \langle x, \mathrm{rank}_A(x) \rangle)) \vee (y \in A \wedge f(x,y) = \langle y, \mathrm{rank}_A(y) \rangle))]$.
2. *P-sr $= \{A \mid A$ is polynomial-time semi-rankable$\}$.*

It turns out that the P-semi-rankable sets are exactly those sets that are both P-selective and weakly P-rankable.

Theorem 6.15   P-sr = P-sel $\cap$ weakly-P-rankable.

So P $\subseteq$ P-sr = P-sel $\cap$ weakly-P-rankable. Can one say in more detail how these classes are related? In fact, the P-sr sets not only differ from P but also cannot even be "approximated from inside" by any P set (in the standard technical terminology, "P-sr is P-immune").

Definition 6.16   *Given any classes C_1 and C_2, we say C_1 is C_2-immune if there is an infinite set in C_1 that has no infinite subset that is a member of C_2.*

Theorem 6.17   P $\subsetneq$ P-sr. *Indeed P-sr is P-immune.*

The class of P-semi-rankable sets forms a proper subset of the P-selective sets and of the weakly P-rankable sets.

Theorem 6.18   P-sr $\subsetneq$ P-sel, *and* P-sr $\subsetneq$ weakly-P-rankable.

Note that in light of Theorem 6.15, Theorem 6.18 is equivalent to stating that P-sel and weakly-P-rankable are incomparable classes—neither contains the other.

We conclude this section with an open research question related to the P-semi-rankable sets. Recall that P $\subseteq$ P-sr $\subseteq$ P-sel. Theorem 6.17 noted that not only does P-sr strictly contain P, but also the stronger result holds that P-sr is P-immune. We also know that P-sr $\subsetneq$ P-sel. However, does an immunity result hold between P-sel and P-sr? That is, can one prove that P-sel is (P-sr)-immune?

This problem remains open. It may be difficult to resolve, since all currently known construction techniques that build P-selective sets can be seen to in fact build no P-selective sets that lack obvious infinite P-sr subsets. So proving that P-sel is (P-sr)-immune via an explicit construction would seem to require new techniques for building P-selective sets. The following result, while neither establishing nor precluding the result "P-sel is (P-sr)-immune," shows that a weaker (in light of Theorem 6.15) version of this statement fails to hold if P = NP.

Theorem 6.19 *Every infinite* P-*selective set has an infinite subset belonging to the class* weakly-FP$^{\Sigma_2^p}$-rankable.

Corollary 6.20 *If* P = NP *then* P-sel *is not* (weakly-P-rankable)-*immune.*

6.3 Associative P-Selectivity: A Potential Refinement of P-Selectivity

And now, at the conclusion of this book, we come full circle. We return to a central topic from the very start of the book: the algebraic properties of P-selector functions. Theorem 1.4 proved that every P-selective set has a *commutative* (i.e., symmetric) P-selector function. We invoked commutativity—via Theorem 1.4 and via its nondeterministic sibling, Theorem 1.24—in the proofs of countless results throughout this book.

In this section, we discuss the natural companion property: associativity. We will see that associativity may be closely linked to a central theme of this book: the advice complexity of the P-selective sets.

As noted above, Theorem 1.4 proves that every P-selective set is in fact P-selective via some *commutative* (i.e., symmetric) P-selector function. What can one say about sets having associative P-selector functions? Does each P-selective set have some associative function under which it is P-selective? What properties do the associatively P-selective sets have?

For clarity, we will formally define associativity. We will do so here just for total, single-valued functions, and we will discuss associativity results only with respect to the P-selective sets. See the Bibliographic Notes of this chapter for references that, using definitions of associativity broad enough to handle partial and multivalued functions, discuss associativity for nondeterministically selective sets.

Definition 6.21 *For any 2-ary total function* $f : \Sigma^* \rightarrow \Sigma^*$, *we say* f *is associative* exactly if

$$(\forall a, b, c \in \Sigma^*)[f(a, f(b, c)) = f(f(a, b), c)].$$

Though the study of associative selectivity started quite recently, much is known. Theorem 6.22 shows that it is at least possible that every P-selective set has some associative function under which it is P-selective. In particular, if P = NP then this certainly is the case.

Theorem 6.22 *Every* P-*selective set has a total, commutative, associative selector function computable in* $\mathrm{FP}^{\mathrm{NP}}$.

It is important to note that, though P = NP is thus a sufficient condition for all P-selective sets to be associatively P-selective, it is not at all clear that P = NP is a necessary condition for this to hold. One might well hope that it is not a necessary condition, since part 1 of Theorem 6.23 below shows a very nice consequence that would hold were all P-selective sets to be associatively P-selective. It would be exciting to reach that consequence without in the process needing to induce P = NP, which itself is so strong as to directly imply the nice consequence.

Theorem 6.23

1. *Every* P-*selective set that has an associative* P-*selector function is in the advice class* $\mathrm{P}/n+1$.
2. *There are associatively* P-*selective sets that are not in* P/n.

That is, from Chapter 2 we know that all P-selective sets are in NP/linear and are in P/quadratic. From that, it is clear that if P = NP then all P-selective sets are in P/linear—the best of both worlds. However, the above theorem gives a potentially different route to concluding that P-sel ⊆ P/linear, namely, if all P-selective sets are associatively P-selective then it follows that P-sel ⊆ P/linear.

Though Theorem 1.4 shows that every P-selective set is P-selective via some commutative (i.e., symmetric) P-selector function, it does not immediately follow that every associatively P-selective set has a commutative, associative P-selector function. The potential problem is that the transformation to achieve commutativity could potentially destroy associativity. Nonetheless, it turns out that one can achieve commutativity in such a way as to preserve associativity. In particular, the commutativity-creating transformation that we used in the proof of Theorem 1.24 can be proven to have this property. Thus, we obtain the following result.

Theorem 6.24 *Every set that is associatively* P-*selective is associatively, commutatively* P-*selective.*

6.4 Bibliographic Notes

Weak selectivity and the results of Section 6.1.1 are due to Ko [Ko83]. Multiselectivity and the results of Section 6.1.2 are due to Hemaspaandra et al. [HJRW97,HJRW98].

Cheatable and P-superterse sets were introduced by Beigel [Bei87]. The recursion-theoretic versions of these notions are due to Beigel, Gasarch, Gill, and Owings [BGGO93]. The notion $(a, b)_p$ verboseness was introduced by Beigel, Kummer, and Stephan [BKS95b]. Near-testability is due to Goldsmith et al. [GJY87,GHJY91]. Hoene and Nickelsen [HN93] introduced the notion "easily-k-countable," and Theorem 6.10 is due to them. The recursion-theoretic version of this notion was studied by Kummer [Kum92]. Hoene and Nickelsen [HN93] defined a precursor of multiselectivity called (A, k)-selectivity and they also defined the following notion of (A, k)-sorting. In this book we frequently use the famous, central fact that a finite subset of a P-selective set can be ordered according to the P-selector such that membership or nonmembership of one element in the sequence implies membership or nonmembership for its successors or predecessors, depending on the way in which the sequence is sorted. An (A, k)-sort function is a function that generates a sorted sequence when the input is a k-tuple of strings. The notion of membership comparability is due to Ogihara [Ogi95], and a restricted version, bounded approximability (bAPP), was studied by Beigel, Kummer, and Stephan [BKS95a]. Recent work of Cai [Cai01] allows the conclusion of Theorem 6.7 to be strengthened to PH $\subseteq$ S$_2$. Theorem 6.8, or versions very close to it, were independently obtained by Agrawal and Arvind [AA96], Beigel, Kummer, and Stephan [BKS95a], and Ogihara [Ogi95]. Wang [Wan95] introduced FPP-selectivity and proved the results of Section 6.1.4.

The class P-sr and the results of Section 6.2 are due to Hemaspaandra, Zaki, and Zimand [HZZ96] and Hemaspaandra et al. [HOZZ].

The notion of associative selector functions and the results of Section 6.3 are due to Hemaspaandra, Hempel, and Nickelsen [HHN01,HHN02]. In this chapter, we discussed only the single-valued, total case, in the context of deterministic selector functions. However, Hemaspaandra, Hempel, and Nickelsen [HHN02] provide many results—some rather different than the deterministic case—on whether sets selective via (potentially partial, potentially multivalued) nondeterministic selector functions are likely to be associatively selective.

A. Definitions of Reductions and Complexity Classes, and Notation List

A.1 Reductions

Below, in naming the reductions, "polynomial-time" is implicit unless otherwise noted, e.g., "many-one reduction" means "many-one polynomial-time reduction."

This list sorts among the $\leq_b^a$ first by b, and in the case of identical b's, by a.

$\leq_\gamma$ – Gamma reduction.
- $A \leq_\gamma B$ if there is a nondeterministic polynomial-time machine N such that (a) for each string x, $N(x)$ has at least one accepting path, and (b) for each string x and each accepting path p of $N(x)$ it holds that: $x \in A$ if and only if the output of $N(x)$ on path p is an element of B.

$\leq_m^p$ – Many-one reduction.
- $A \leq_m^p B$ if $(\exists f \in \mathrm{FP})(\forall x)[x \in A \iff f(x) \in B]$.

$\leq_{pos}^p$ – (Globally) Positive Turing reduction.
- $A \leq_{pos}^p B$ if there is a deterministic Turing machine M and a polynomial q such that
  1. $(\forall D)[\mathrm{runtime}_{M^D}(x) \leq q(|x|)]$,
  2. $A = L(M^B)$, and
  3. $(\forall C, D)[C \subseteq D \implies L(M^C) \subseteq L(M^D)]$.

$\leq_{\widehat{pos}}^p$ – Locally positive Turing reduction.
- $A \leq_{\widehat{pos}}^p B$ if there is a deterministic polynomial-time Turing machine M such that
  1. $A = L(M^B)$,
  2. $(\forall C)[L(M^{B \cup C}) \supseteq L(M^B)]$, and
  3. $(\forall C)[L(M^{B-C}) \subseteq L(M^B)]$.

$\leq_T^e$ – Exponential-time Turing reduction.
- $A \leq_T^e B$ if $A \in \mathrm{E}^B$. If $A \in \mathrm{E}^B$ via an exponential-time machine, M, having the property that on each input of length n it holds that M makes at most $g(n)$ queries to its oracle, we write $A \leq_{g(n)\text{-}T}^e B$. If for some polynomial $g(n)$ it holds that $A \leq_{g(n)\text{-}T}^e B$, we say $A \leq_{\mathrm{poly}\text{-}T}^e B$.

$\leq_T^p$ – Turing reduction.
- $A \leq_T^p B$ if $A \in \mathrm{P}^B$.

$\leq_T^{sn}$ – Strong nondeterministic Turing reduction.
  – $A \leq_T^{sn} B$ if $A \in \mathrm{NP}^B \cap \mathrm{coNP}^B$.

$\leq_{tt}^p, \leq_{ptt}^p, \leq_{btt}^p, \leq_{k\text{-}tt}^p$
  – Various types of truth-table reductions.
  – Truth-table reductions are often defined in terms of separate "query generation" and "answer interpretation" machines. In this book, we will more commonly use a more intuitive, though equivalent, formulation. We say that $A \leq_{tt}^p B$ if there is a Turing machine, having a designated query tape, that on each input x will enter a special query state at most once. When it does, our model is that the content of the query tape is interpreted as a collection of strings (say, each separated by the character "#"), and in one time step the content of the query tape is replaced by the bitstring $b_1 b_2 \cdots b_z$, where b_i is 1 if the ith string on the tape is an element of B and is 0 otherwise, and where z is the number of strings that were on the query tape. The Turing machine in this same step has its state moved to a special query-was-just-answered state. (It is henceforth not allowed to query.) It then continues on, eventually accepting or rejecting. Note in particular that the query tape is readable, and so our Turing machine may read the bitstring $b_1 b_2 \cdots b_z$, and these bits may affect the acceptance.

  For each natural number k, if $A \leq_{tt}^p B$ via a Turing machine M of the form described above then we say that $A \leq_{k\text{-}tt}^p B$ if on no input does the machine ever enter the query state with more than $k-1$ "#" characters on its query tape. If there is a natural number k such that $A \leq_{k\text{-}tt}^p B$, then we say that $A \leq_{btt}^p B$. The reducibility $\leq_{btt}^p$ is referred to as bounded-truth-table reducibility.

  Suppose that $A \leq_{tt}^p B$ via a Turing machine M of the form described in the above definition of $\leq_{tt}^p$. We will, when this use is clear from content (i.e., when the machine is a truth-table machine), use M^B to denote the computation of M when B is used to answer the query list written to the oracle tape. We say that $A \leq_{ptt}^p B$ (A positive-truth-table reduces to B) if there exists a polynomial-time Turing machine M (that functions as a truth-table machine) such that $A \leq_{tt}^p B$ via M, and $(\forall C, D)[L(M^C) \subseteq L(M^{C \cup D})]$.

A.2 Complexity Classes

This list is, with some exceptions for clarity, alphabetically "word-ordered." Also, it orders Greek letters under their Romanized versions (e.g., Σ's are alphabetized as if they were "Sigma"s).

$\mathrm{co} \cdot \mathcal{C}$
  – Set-wise complements of complexity class $\mathcal{C}$.

- $A \in \mathrm{co} \cdot \mathcal{C}$ if $\overline{A} \in \mathcal{C}$.

$\mathcal{C}\text{-}\leq_r\text{-complete}$
- Let $\leq_r$ be any reducibility and let $\mathcal{C}$ be any complexity class. A set A is said to be $\mathcal{C}\text{-}\leq_r\text{-complete}$ if $A \in \mathcal{C}$ and A is $\mathcal{C}\text{-}\leq_r\text{-hard}$.

$\mathcal{C}\text{-}\leq_r\text{-hard}$
- Let $\leq_r$ be any reducibility and let $\mathcal{C}$ be any complexity class. A set A is said to be $\mathcal{C}\text{-}\leq_r\text{-hard}$ if, for each $B \in \mathcal{C}$, $B \leq_r A$.

$\mathcal{C}\text{-complete}$
- Let $\mathcal{C}$ be any complexity class. A set is said to be $\mathcal{C}$-complete if it is $\mathcal{C}\text{-}\leq_m^p\text{-complete}$.

$\mathcal{C}$-hard – Let $\mathcal{C}$ be any complexity class. A set is said to be $\mathcal{C}$-hard if it is $\mathcal{C}\text{-}\leq_m^p\text{-hard}$.

$\mathcal{C}/\{\mathcal{F}\}$
- A token-based advice class.
- Let $\mathcal{F}$ be any collection of (total) functions from $\mathbb{N}$ to $\mathbb{N}^+$. Let $\mathcal{C}$ be any collection of sets. Define

$$\mathcal{C}/\{\mathcal{F}\} = \{A \mid (\exists g \in \mathcal{F})[A \in \mathcal{C}/\{g\}]\}.$$

$\mathcal{C}/\{f\}$
- A token-based advice class.
- Let $f : \mathbb{N} \to \mathbb{N}^+$. Assume that natural numbers have their standard encoding over binary strings. Let $\mathcal{C}$ be any collection of sets. Define
$$\mathcal{C}/\{f\} = \{A \mid (\exists B \in \mathcal{C})(\exists h : \mathbb{N} \to \mathbb{N}^+)$$
$$[(\forall n)[h(n) \in \{1, \ldots, f(n)\}] \text{ and } (\forall x \in \Sigma^*)[x \in A \iff \langle x, h(|x|)\rangle \in B]]\}.$$

$\mathcal{C}/\mathcal{F}$ – A length-based advice class.
- Let $\mathcal{F}$ be any class of (total) functions mapping from $\mathbb{N}$ to $\mathbb{N}$. Define

$$\mathcal{C}/\mathcal{F} = \{A \mid (\exists f \in \mathcal{F})[A \in \mathcal{C}/f]\}.$$

- One common value of $\mathcal{F}$ is "poly," which denotes the class of polynomials.

$\mathcal{C}/f$ – A length-based advice class.
- Let $f : \mathbb{N} \to \mathbb{N}$ be any (total) function. Let $\mathcal{C}$ be any collection of sets. Define $\mathcal{C}/f = \{A \mid (\exists B \in \mathcal{C})(\exists h : \mathbb{N} \to \{0,1\}^*)[(\forall n)[|h(n)| = f(n)] \wedge (\forall x \in \Sigma^*)[x \in A \iff \langle x, h(|x|)\rangle \in B]]\}$.

coNP – Co-nondeterministic polynomial time.
- $A \in \mathrm{coNP}$ if $\overline{A} \in \mathrm{NP}$.

Δ_k^p, $k \geq 0$
- The kth "Δ" level of the polynomial hierarchy.
- $\Delta_k^p = \Delta_k^{p,\emptyset}$.

$\Delta_k^{p,A}$, $k \geq 0$
- The kth "Δ" level of the polynomial hierarchy relativized via oracle A.
- $\Delta_0^{p,A} = \mathrm{P}^A$.

- For $k \geq 1$, $\Delta_k^{p,A} = P^{\Sigma_{k-1}^{p,A}}$.
- See, however, the discussion in footnote 6 (on page 43), since the analog of that holds here.

Disjunctively self-reducible sets
- A set A is said to be *disjunctively self-reducible* if there is a deterministic polynomial-time Turing machine M such that
    1. $A = L(M^A)$,
    2. for each x, $M^A(x)$ queries only strings of lengths strictly less than $|x|$, and
    3. the acceptance behavior of M is such that on each input M accepts exactly when either (a) M asks at least one query that is in the oracle set, or (b) M asks no queries and halts in an accepting state.

    If A is disjunctively self-reducible via a machine M that on each input asks at most two oracle questions, then we say that A is *2-disjunctively self-reducible*.

DSPACE$[f(n)]$
- Deterministic space $f(n)$.
- $A \in \text{DSPACE}[f(n)]$ if A is accepted by a deterministic Turing machine whose running space is $\mathcal{O}(f(n))$.

DTIME$[f(n)]$
- Deterministic time $f(n)$.
- $A \in \text{DTIME}[f(n)]$ if A is accepted by a deterministic Turing machine whose running time is $\mathcal{O}(f(n))$.

E   - Deterministic exponential time.
- $E = \bigcup_{k>0} \text{DTIME}[2^{kn}]$.

$\text{EL}_{\Delta_k^p}$, $k \geq 2$
- The Δ_k^p extended low sets.
- $A \in \text{EL}_{\Delta_k^p}$ if $\Delta_k^{p,A} \subseteq \Delta_{k-1}^{p,A\oplus\text{SAT}}$.

$\text{EL}_{\Delta_k^p}^W$, $k \geq 2$
- The Δ_k^p extended low sets in relativized world W.
- $A \in \text{EL}_{\Delta_k^p}^W$ if $\Delta_k^{p,A\oplus W} \subseteq \Delta_{k-1}^{p,A\oplus\text{SAT}\oplus W}$.

ELH - The extended low hierarchy.
- $\text{ELH} = \bigcup_{k\geq2} \text{EL}_{\Sigma_k^p}$.

$\text{EL}_{\Sigma_k^p}$, $k \geq 2$
- The Σ_k^p extended low sets.
- $A \in \text{EL}_{\Sigma_k^p}$ if $\Sigma_k^{p,A} \subseteq \Sigma_{k-1}^{p,A\oplus\text{SAT}}$.

$\text{EL}_{\Theta_k^p}$, $k \geq 2$
- The Θ_k^p extended low sets.
- $A \in \text{EL}_{\Theta_k^p}$ if $\Theta_k^{p,A} \subseteq \Theta_{k-1}^{p,A\oplus\text{SAT}}$.

$\text{E}_r(\mathcal{C})$
- The sets $\leq_r$-equivalent to some set in $\mathcal{C}$.
- $A \in \text{E}_r(\mathcal{C})$ if $(\exists B \in \mathcal{C})[A \leq_r B \wedge B \leq_r A]$.
- Note: $\leq_r$ must be a defined reduction type.

EXP – Deterministic "polynomial exponential" time.
- $EXP = \bigcup_{k>0} DTIME[2^{n^k}]$.

$\mathcal{F}$-sel
- Selectivity via general functions.
- Let $\mathcal{F}$ be a class of functions. We say a set A is $\mathcal{F}$-selective if there is an $f \in \mathcal{F}$ such that, for each x and y,
  1. set-$f(x, y) \subseteq \{x, y\}$, and
  2. if $A \cap \{x, y\} \neq \emptyset$ then $\emptyset \neq$ set-$f(x, y) \subseteq A$.
  We say such a function f is an $\mathcal{F}$-selector for A. Let $\mathcal{F}$ be a class of functions. $\mathcal{F}$-sel denotes $\{A \mid A \text{ is } \mathcal{F}\text{-selective}\}$.

FewP – Polynomial-ambiguity (nondeterministic) polynomial time.
- $B \in$ FewP if $(\exists$ polynomial-time 2-ary predicate $R)(\exists$ polynomial $q)$ $(\exists$ polynomial $r)(\forall x)[(\|\{z \mid |z| \leq q(|x|) \wedge R(x, z)\}\| \leq r(|x|)) \wedge (x \in B \iff (\exists y)[|y| \leq q(|x|) \wedge R(x, y)])]$.

FP – Deterministic polynomial-time computable functions.
- $f \in$ FP if f is a (total, single-valued) function computable by a deterministic polynomial-time Turing machine.
- All function classes are implicitly partial (unless subscripted with a "t" to denote totality) *except* FP, which by longstanding convention represents the total functions that are computable in deterministic polynomial time.

HH – The high hierarchy.
- $HH = \bigcup_{k \geq 0} H_{\Sigma_k^p}$.

$H_{\Sigma_k^p}$, $k \geq 0$ – The Σ_k^p-high sets.
- $A \in H_{\Sigma_k^p}$ if $A \in NP$ and $\Sigma_k^{p,A} \supseteq \Sigma_{k+1}^p$.

$L_{\mathcal{C}}$ – The $\mathcal{C}$-low sets.
- $A \in L_{\mathcal{C}}$ if $A \in NP$ and $\mathcal{C}^A = \mathcal{C}$.
- Note: $\mathcal{C}$ must be a class for which relativization has been defined.

$L_{\mathcal{C}}^W$ – $\mathcal{C}$-low sets in oracle world W.
- $A \in L_{\mathcal{C}}^W$ if $A \in NP^W$ and $\mathcal{C}^{A \oplus W} = \mathcal{C}^W$.
- Note: $\mathcal{C}$ must be a class for which relativization has been defined.

LH – The low hierarchy.
- $LH = \bigcup_{k \geq 0} L_{\Sigma_k^p}$.

NE – Nondeterministic exponential time.
- $NE = \bigcup_{k>0} NTIME[2^{kn}]$.

NEXP – Nondeterministic "polynomial exponential" time.
- $NEXP = \bigcup_{k>0} NTIME[2^{n^k}]$.

NNT – The implicitly membership-testable sets (also known as the nearly near-testable sets).
- $A \in$ NNT if $(\exists f \in FP)(\forall x)[(f(x) = \text{"in"} \wedge x \in A) \vee (f(x) = \text{"out"} \wedge x \notin A) \vee (f(x) = \text{"xor"} \wedge \|\{x, predecessor(x)\} \cap A\| = 1) \vee (f(x) = \text{"nxor"} \wedge \|\{x, predecessor(x)\} \cap A\| \equiv 0 \pmod 2))]$.

NP – Nondeterministic polynomial time.
- $NP = \bigcup_{k>0} NTIME[n^k]$.

- $B \in \mathrm{NP}$ if ($\exists$ polynomial-time 2-ary predicate R)($\exists$ polynomial q)($\forall x$) $[x \in B \iff (\exists y)[|y| \leq q(|x|) \wedge R(x,y)]]$.

(NP $\cap$ coNP)/poly
- See entry for $\mathcal{C}/\mathcal{F}$.

NP/poly
- See entry for $\mathcal{C}/\mathcal{F}$.

NP^A – Nondeterministic polynomial time relative to oracle A.
- The class of languages accepted by nondeterministic polynomial-time machines given unit-cost access to oracle A.
- $B \in \mathrm{NP}^A$ if there is a 2-ary predicate R—computable in polynomial time relative to A (i.e., being informal about the type of the class P^A, $R \in \mathrm{P}^A$)—and there is a polynomial q such that: ($\forall x$)[$x \in B \iff (\exists y)$ $[|y| \leq q(|x|) \wedge R(x,y)]]$.

NPMV
- The multivalued nondeterministic polynomial-time functions.
- Each nondeterministic polynomial-time Turing machine is considered to be a function-computing machine as follows. Each path that rejects is considered to have no output. Each path that accepts is considered to output the string of characters stretching, at the moment that path accepts, from the left end of its semi-infinite worktape through (but not including) the character underneath its worktape head. NPMV denotes the class of all functions f that can be computed (in the sense just stated) by some nondeterministic polynomial-time Turing machine.

NPMV-sel
- See entries for $\mathcal{F}$-sel and NPMV.

NPMV$_t$
- The total, multivalued, nondeterministic polynomial-time functions.
- $f \in \mathrm{NPMV}_t$ if f is total (i.e., for all x and y, $\|\text{set-}f(x,y)\| > 0$) and $f \in \mathrm{NPMV}$.

NPMV$_t$-sel
- See entries for $\mathcal{F}$-sel and NPMV$_t$.

NPSV
- The single-valued nondeterministic polynomial-time functions.
- $f \in \mathrm{NPSV}$ if f is single-valued (i.e., for all x and y, $\|\text{set-}f(x,y)\| \leq 1$) and $f \in \mathrm{NPMV}$.

NPSV-sel
- See entries for $\mathcal{F}$-sel and NPSV.

NPSV$_t$
- The total, single-valued, nondeterministic polynomial-time functions.
- $f \in \mathrm{NPSV}_t$ if f is total and $f \in \mathrm{NPSV}$.

NPSV$_t$-sel
- See entries for $\mathcal{F}$-sel and NPSV$_t$.

NT – The near-testable sets.
- $A \in \mathrm{NT}$ if $(\exists f \in \mathrm{FP})(\forall x)[$
  $(f(x) = \text{``}xor\text{''} \wedge \|\{x, predecessor(x)\} \cap A\| = 1) \vee$
  $(f(x) = \text{``}nxor\text{''} \wedge \|\{x, predecessor(x)\} \cap A\| \equiv 0 \pmod 2))]$.

NTIME$[f(n)]$
- Nondeterministic time $f(n)$.
- $A \in \mathrm{NTIME}[f(n)]$ if A is accepted by a nondeterministic Turing machine whose nondeterministic running time is $\mathcal{O}(f(n))$.

P – Deterministic polynomial time.
- $\mathrm{P} = \bigcup_{k>0} \mathrm{DTIME}[n^k]$.

P-close
- The P-close sets.
- $A \in \text{P-close}$ if $(\exists B \in \mathrm{P})(\exists S \in \mathrm{SPARSE})[A = B \triangle S]$, where $B \triangle S = (B - S) \cup (S - B)$.

P-sel – The P-selective sets; the semi-feasible sets.
- $A \in \text{P-sel}$ if $(\exists f \in \mathrm{FP})(\forall x, y)[f(x, y) \in \{x, y\} \wedge (f(x, y) \cap A \neq \emptyset \implies f(x, y) \in A)]$. Such a function f is called a P-*selector function* for A.

P/poly
- See entry for $\mathcal{C}/\mathcal{F}$.

P^A – Deterministic polynomial time relative to oracle A.
- The class of languages accepted by deterministic polynomial-time machines given unit-cost access to oracle A.

$\mathrm{P}^{A[f(n)]}$
- Deterministic polynomial time relative to oracle A, with a bounded number of queries.
- The class of languages accepted by deterministic polynomial-time machines given unit-cost access to oracle A and allowed on each input x at most $f(|x|)$ oracle queries.
- The obvious generalizations of this from a single oracle to a class of oracles, or from a single function to a class of functions, or both (e.g., $\mathrm{P}^{\mathcal{C}[\mathcal{F}]}$) are defined and used in the obvious analogous ways.

PH – The polynomial hierarchy.
- $\mathrm{PH} = \bigcup_{k \geq 0} \Sigma_k^p = \mathrm{P} \cup \mathrm{NP} \cup \mathrm{NP}^{\mathrm{NP}} \cup \mathrm{NP}^{\mathrm{NP}^{\mathrm{NP}}} \cup \cdots$.
- Note: We say that the polynomial hierarchy *collapses* if $(\exists k)[\mathrm{PH} = \Sigma_k^p]$.

PP – (Unbounded error) Probabilistic polynomial time.
- $L \in \mathrm{PP}$ if there is a nondeterministic polynomial-time Turing machine M such that, for every input x, it holds that $x \in L$ if and only if more than half of the computation paths of $M(x)$ are accepting paths.

PSPACE
- Polynomial space.
- $\mathrm{PSPACE} = \bigcup_{k>0} \mathrm{DSPACE}[n^k]$.

qP – Quasipolynomial time.
- $\mathrm{qP} = \bigcup_{k \geq 0} \mathrm{DTIME}[2^{\log^k n}]$.

R   – Random polynomial time.

– $B \in R$ if $(\exists$ polynomial-time 2-ary predicate $R)(\exists$ polynomial $q)(\forall x)$
$[(x \notin B \implies \|\{z \mid |z| \leq q(|x|) \wedge R(x,z)\}\| = 0) \wedge (x \in B \implies$
$\|\{z \mid |z| \leq q(|x|) \wedge R(x,z)\}\| / \|\{z \mid |z| \leq q(|x|)\}\| \geq 1/2)]$.

$R_r(\mathcal{C})$

– The sets that $\leq_r$-reduce to some set in $\mathcal{C}$.

– $A \in R_r(\mathcal{C})$ if $(\exists B \in \mathcal{C})[A \leq_r B]$.

– Note: $\leq_r$ must be a defined reduction type.

S_2 – The second level of the symmetric alternation hierarchy.

– $A \in S_2$ if there exists a set $R \in P$ and a polynomial q such that, for every x,

1. if $x \in A$ then $(\exists y : |y| \leq q(|x|))(\forall z : |z| \leq q(|x|))[\langle x, y, z \rangle \in R]$, and

2. if $x \notin A$ then $(\exists z : |z| \leq q(|x|))(\forall y : |y| \leq q(|x|))[\langle x, y, z \rangle \notin R]$.

– It is known that $BPP \cup \Delta_2^p \subseteq S_2 \subseteq ZPP^{NP} \subseteq NP^{NP}$.

$S_2^{NP \cap coNP}$ – The second level of the symmetric alternation hierarchy relativized to $NP \cap coNP$.

– $A \in S_2^{NP \cap coNP}$ if there exists a set $R \in P^{NP \cap coNP}$ (equivalently, $R \in NP \cap coNP$) and a polynomial q such that, for every x,

1. if $x \in A$ then $(\exists y : |y| \leq q(|x|))(\forall z : |z| \leq q(|x|))[\langle x, y, z \rangle \in R]$, and

2. if $x \notin A$ then $(\exists z : |z| \leq q(|x|))(\forall y : |y| \leq q(|x|))[\langle x, y, z \rangle \notin R]$.

– It is known that $BPP \cup \Delta_2^p \subseteq S_2 \subseteq S_2^{NP \cap coNP} \subseteq ZPP^{NP} \subseteq NP^{NP}$.

Semi-recursive sets

– We say a set A is semi-recursive if $(\exists$ recursive function $f)(\forall x, y)$
$[(f(x,y) = x \vee f(x,y) = y) \wedge (f(x,y) \cap A \neq \emptyset \implies f(x,y) \in A)]$.

set-f

– Outputs of a multivalued function.

– Let f be any (possibly partial, possibly multivalued) function. For any strings x and y, set-$f(x,y)$ denotes $\{z \mid z$ is an output of $f(x,y)\}$.

Σ_k^0 – The kth level of the arithmetical hierarchy (also known as the Kleene Hierarchy).

Σ_k^p, $k \geq 0$

– The kth "Σ" level of the polynomial hierarchy.

– $\Sigma_k^p = \Sigma_k^{p,\emptyset}$.

$\Sigma_k^{p,A}$, $k \geq 0$

– The kth "Σ" level of the polynomial hierarchy relativized via oracle A.

– $\Sigma_0^{p,A} = P^A$.

– For $k \geq 1$, $\Sigma_k^{p,A} = NP^{\Sigma_{k-1}^{p,A}}$.

SPARSE

– The sparse sets.

– $A \in SPARSE$ if $(\exists$ polynomial $q)(\forall n)[\|A^{=n}\| \leq q(n)]$.

TALLY

– The tally sets.

– $A \in TALLY$ if $A \subseteq \{\epsilon, 1, 11, 111, \ldots\}$.

Θ_k^p, $k \geq 0$

    – The kth "Θ" level of the polynomial hierarchy.

    – $\Theta_k^p = \Theta_k^{p,\emptyset}$.

$\Theta_k^{p,A}$, $k \geq 0$

    – The kth "Θ" level of the polynomial hierarchy relativized via oracle A.

    – $\Theta_0^{p,A} = P^A$.

    – For $k \geq 1$, $\Theta_k^{p,A} = P^{\Sigma_{k-1}^{p,A}[\mathcal{O}(\log n)]}$, where as usual $P^{\mathcal{C}[\mathcal{F}]}$ denotes the union over all sets A in $\mathcal{C}$ and all functions f in $\mathcal{F}$ of the class of languages acceptable by P machines with oracle A that on each input x make at most $f(|x|)$ oracle queries.

    – See, however, the discussion in footnote 6 (on page 43).

Turing self-reducible sets

    – A set A is *Turing self-reducible* if there is a deterministic polynomial-time Turing machine M such that $A = L(M^A)$ and, for each x, $M^A(x)$ queries only strings of lengths strictly less than $|x|$.

UP – Unambiguous (nondeterministic) polynomial time.

    – $B \in \text{UP}$ if ($\exists$ polynomial-time 2-ary predicate R)($\exists$ polynomial q) $(\forall x)[(\|\{z \mid |z| \leq q(|x|) \wedge R(x,z)\}\| \leq 1) \wedge (x \in B \iff (\exists y)[|y| \leq q(|x|) \wedge R(x,y)])]$.

    – We say a Turing machine N is unambiguous if and only if, for all x, it holds that N on input x has at most one accepting computation path. A set is in UP exactly if it is accepted by some polynomial-time, unambiguous Turing.

ZPP

    – Expected polynomial time.

    – A set is in ZPP if there is a probabilistic Turing machine that accepts the set (without error) and whose expected running time is polynomially bounded in the length of the input.

    – It is known that ZPP $= \text{R} \cap \text{coR}$.

A.3 Some Other Notation

ϵ	The empty string.
$a \in A$	a is a member of set A.
$a \notin A$	a is not a member of set A.
$=, \neq, \leq, \geq, <, >$	Standard arithmetic relations.
$=, \subseteq, \supseteq, \not\subseteq, \not\supseteq, \subsetneq, \supsetneq$	Standard set relations.
$\emptyset$	The empty set.
A^n	$\{x \mid (\exists z_1, \ldots, z_n \in A)[x = z_1 z_2 \cdots z_n]\}$.
$A^{=n}$	$\{x \mid x \in A \wedge |x| = n\}$.
$A^{\leq n}$	$\{x \mid x \in A \wedge |x| \leq n\}$.
$A^{<n}$	$\{x \mid x \in A \wedge |x| < n\}$.
$|a|$	Length of string a.

$A \triangle B$	$(A - B) \cup (B - A)$.
$\overline{A}$	The complement of A: $\Sigma^* - A$.
$A \cup B$	$\{x \mid x \in A \vee x \in B\}$.
$a \wedge b$	Logical "and" of the two boolean variables.
$a \vee b$	Logical "or" the two boolean variables.
$\neg$	Logical negation of a boolean value.
$\|A\|$	The cardinality of set A.
$A - B$	$A \cap \overline{B}$.
$A \cap B$	$\{x \mid x \in A \wedge x \in B\}$.
$A \oplus B$	$\{0x \mid x \in A\} \cup \{1y \mid y \in B\}$.
$\mathbb{N}$	$\{0, 1, 2, \ldots\}$.
$\mathbb{N}^+$	$\{1, 2, 3, \ldots\}$.
Σ	The input alphabet, which unless otherwise stated we assume consists of at least two characters, 0 and 1. Unless otherwise stated, sets are subsets of Σ^*.
A/R	The set of equivalence classes of A with respect to equivalence relation R.
$cl(a)$	The equivalence class containing a (with respect to a set of equivalence classes that is implicit from the context).
$\leq_f$	When f is a symmetric P-selector function, we will sometimes use $a \leq_f b$ to denote $f(a, b) = b$.
$\prec$	Used to denote a linear ordering.
$\mathcal{O}(f(n))$	Let $g : \mathbb{N} \to \mathbb{N}$ and $f : \mathbb{N} \to \mathbb{N}$. We say $g(n) = \mathcal{O}(f(n))$ if $(\exists c > 0)(\exists n_0 \geq 0)(\forall n)[n \geq n_0 \implies g(n) \leq cf(n)]$.

References

[AA96]   M. Agrawal and V. Arvind. Quasi-linear truth-table reductions to P-selective sets. *Theoretical Computer Science*, 158(1–2):361–370, 1996.

[ABG90]   A. Amir, R. Beigel, and W. Gasarch. Some connections between bounded query classes and non-uniform complexity. In *Proceedings of the 5th Structure in Complexity Theory Conference*, pages 232–243. IEEE Computer Society Press, July 1990.

[AFK89]   M. Abadi, J. Feigenbaum, and J. Kilian. On hiding information from an oracle. *Journal of Computer and System Sciences*, 39(1):21–50, 1989.

[AH92]   E. Allender and L. Hemachandra. Lower bounds for the low hierarchy. *Journal of the ACM*, 39(1):234–251, 1992.

[All90]   E. Allender. Oracles versus proof techniques that do not relativize. In *Proceedings of the 1990 SIGAL International Symposium on Algorithms*, pages 39–52. Springer-Verlag *Lecture Notes in Computer Science #450*, August 1990.

[ALM+98]   S. Arora, C. Lund, R. Motwani, M. Sudan, and M. Szegedy. Proof verification and the hardness of approximation problems. *Journal of the ACM*, 45(3):501–555, 1998.

[Bar89]   D. Barrington. Bounded-width polynomial-size branching programs recognize exactly those languages in NC^1. *Journal of Computer and System Sciences*, 38(1):150–164, 1989.

[Bar92]   D. Barrington. Quasipolynomial size circuit classes. In *Proceedings of the 7th Structure in Complexity Theory Conference*, pages 86–93. IEEE Computer Society Press, June 1992.

[Bar95]   D. Barrington, July 19, 1995. Personal communication.

[BB86]   J. Balcázar and R. Book. Sets with small generalized Kolmogorov complexity. *Acta Informatica*, 23(6):679–688, 1986.

[BBS86]   J. Balcázar, R. Book, and U. Schöning. Sparse sets, lowness and highness. *SIAM Journal on Computing*, 15(3):739–746, 1986.

[BC93]   D. Bovet and P. Crescenzi. *Introduction to the Theory of Complexity*. Prentice-Hall, 1993.

[BDG90]   J. Balcázar, J. Díaz, and J. Gabarró. *Structural Complexity II*. EATCS Monographs in Theoretical Computer Science. Springer-Verlag, 1990.

[BDG95]   J. Balcázar, J. Díaz, and J. Gabarró. *Structural Complexity I*. EATCS Texts in Theoretical Computer Science. Springer-Verlag, 2nd edition, 1995.

[Bei87]   R. Beigel. Query-limited reducibilities. Technical Report 87-05, Department of Computer Science, Johns Hopkins University, Baltimore, MD, 1987.

[Bei88]   R. Beigel. NP-hard sets are P-superterse unless R=NP. Technical
             Report 88-04, Department of Computer Science, Johns Hopkins Uni-
             versity, Baltimore, MD, August 1988.
[BFL91]   L. Babai, L. Fortnow, and C. Lund. Non-deterministic exponential
             time has two-prover interactive protocols. *Computational Complexity*,
             1(1):3–40, 1991.
[BFT98]   H. Buhrman, L. Fortnow, and T. Thierauf. Nonrelativizing separations.
             In *Proceedings of the 13th Annual IEEE Conference on Computational
             Complexity*, pages 8–12. IEEE Computer Society Press, June 1998.
[BGGO93]   R. Beigel, W. Gasarch, J. Gill, and J. Owings. Terse, superterse, and
             verbose sets. *Information and Computation*, 103(1):68–85, 1993.
[BI97]   D. Barrington and N. Immerman. Time, hardware, and uniformity. In
             L. Hemaspaandra and A. Selman, editors, *Complexity Theory Retro-
             spective II*, pages 1–22. Springer-Verlag, 1997.
[BKS95a]   R. Beigel, M. Kummer, and F. Stephan. Approximable sets. *Informa-
             tion and Computation*, 120(2):304–314, 1995.
[BKS95b]   R. Beigel, M. Kummer, and F. Stephan. Quantifying the amount of
             verboseness. *Information and Computation*, 118(1):73–90, 1995.
[BL97]   H. Burtschick and W. Lindner. On sets Turing reducible to P-selective
             sets. *Theory of Computing Systems*, 30(2):135–143, 1997.
[BLS84]   R. Book, T. Long, and A. Selman. Quantitative relativizations of
             complexity classes. *SIAM Journal on Computing*, 13(3):461–487, 1984.
[BLS85]   R. Book, T. Long, and A. Selman. Qualitative relativizations of com-
             plexity classes. *Journal of Computer and System Sciences*, 30(3):395–
             413, 1985.
[Boo87]   R. Book. Towards a theory of relativizations: Positive relativizations.
             In *Proceedings of the 4th Annual Symposium on Theoretical Aspects
             of Computer Science*, pages 1–21. Springer-Verlag *Lecture Notes in
             Computer Science #247*, 1987.
[Boo89]   R. Book. Restricted relativizations of complexity classes. In J. Hart-
             manis, editor, *Computational Complexity Theory*, pages 47–74. Amer-
             ican Mathematical Society, 1989. Proceedings of Symposia in Applied
             Mathematics #38.
[BT96]   H. Buhrman and L. Torenvliet. P-selective self-reducible sets: A new
             characterization of P. *Journal of Computer and System Sciences*,
             53(2):210–217, 1996.
[BTvEB93]   H. Buhrman, L. Torenvliet, and P. van Emde Boas. Twenty questions
             to a P-selector. *Information Processing Letters*, 48(4):201–204, 1993.
[Cai01]   J. Cai. $S_2^p \subseteq ZPP^{NP}$. In *Proceedings of the 42nd IEEE Symposium
             on Foundations of Computer Science*, pages 620–629. IEEE Computer
             Society Press, October 2001.
[CCHO01]   J. Cai, V. Chakaravarthy, L. Hemaspaandra, and M. Ogihara. Some
             Karp–Lipton-type theorems based on S_2. Technical Report TR-759,
             Department of Computer Science, University of Rochester, Rochester,
             NY, September 2001. Revised, November 2001.
[CF91]   J. Cai and M. Furst. PSPACE survives constant-width bottlenecks.
             International Journal of Foundations of Computer Science, 2(1):67–
             76, 1991.
[CGH+88]   J. Cai, T. Gundermann, J. Hartmanis, L. Hemachandra, V. Sewelson,
             K. Wagner, and G. Wechsung. The boolean hierarchy I: Structural
             properties. *SIAM Journal on Computing*, 17(6):1232–1252, 1988.
[CH89]   J. Cai and L. Hemachandra. Enumerative counting is hard. *Informa-
             tion and Computation*, 82(1):34–44, 1989.

[CH91]   J. Cai and L. Hemachandra. A note on enumerative counting. *Information Processing Letters*, 38(4):215–219, 1991.

[CHV93]   J. Cai, L. Hemachandra, and J. Vyskoč. Promises and fault-tolerant database access. In K. Ambos-Spies, S. Homer, and U. Schöning, editors, *Complexity Theory*, pages 101–146. Cambridge University Press, 1993.

[DK00]   D. Du and K. Ko. *Theory of Computational Complexity*. John Wiley and Sons, 2000.

[For94]   L. Fortnow. The role of relativization in complexity theory. *Bulletin of the EATCS*, 52:229–244, 1994.

[Gav95]   R. Gavaldà. Bounding the complexity of advice functions. *Journal of Computer and System Sciences*, 50(3):468–475, 1995.

[GH00]   C. Glaßer and L. Hemaspaandra. A moment of perfect clarity I: The parallel census technique. *SIGACT News*, 31(3):37–42, 2000.

[GHJY91]   J. Goldsmith, L. Hemachandra, D. Joseph, and P. Young. Near-testable sets. *SIAM Journal on Computing*, 20(3):506–523, 1991.

[GJY87]   J. Goldsmith, D. Joseph, and P. Young. Self-reducible, P-selective, near-testable, and P-cheatable sets: The effect of internal structure on the complexity of a set. In *Proceedings of the 2nd Structure in Complexity Theory Conference*, pages 50–59. IEEE Computer Society Press, June 1987.

[HCC⁺92]   J. Hartmanis, R. Chang, S. Chari, D. Ranjan, and P. Rohatgi. Relativization: A revisionistic retrospective. *Bulletin of the EATCS*, 47:144–153, 1992.

[Hem93]   L. Hemaspaandra. Lowness: A yardstick for NP−P. *SIGACT News*, 24 (Spring)(2):10–14, 1993.

[HH91]   L. Hemachandra and A. Hoene. On sets with efficient implicit membership tests. *SIAM Journal on Computing*, 20(6):1148–1156, 1991.

[HHN⁺93]   L. Hemaspaandra, A. Hoene, A. Naik, M. Ogiwara, A. Selman, T. Thierauf, and J. Wang. Selectivity: Reductions, nondeterminism, and function classes. Technical Report TR-469, Department of Computer Science, University of Rochester, Rochester, NY, August 1993.

[HHN⁺95]   L. Hemaspaandra, A. Hoene, A. Naik, M. Ogiwara, A. Selman, T. Thierauf, and J. Wang. Nondeterministically selective sets. *International Journal of Foundations of Computer Science*, 6(4):403–416, 1995.

[HHN01]   L. Hemaspaandra, H. Hempel, and A. Nickelsen. Algebraic properties for deterministic selectivity. In *Proceedings of the 4th Annual International Computing and Combinatorics Conference*, pages 49–58. Springer-Verlag *Lecture Notes in Computer Science #2108*, August 2001.

[HHN02]   L. Hemaspaandra, H. Hempel, and A. Nickelsen. Algebraic properties for deterministic and nondeterministic selectivity. Technical Report TR-778, Department of Computer Science, University of Rochester, Rochester, NY, May 2002.

[HHO⁺93]   L. Hemaspaandra, A. Hoene, M. Ogiwara, A. Selman, T. Thierauf, and J. Wang. Selectivity. In *Proceedings of the 5th International Conference on Computing and Information*, pages 55–59. IEEE Computer Society Press, 1993.

[HHO96]   L. Hemaspaandra, A. Hoene, and M. Ogihara. Reducibility classes of P-selective sets. *Theoretical Computer Science*, 155(2):447–457, 1996. Erratum appears in the same journal, 234(1–2):323.

[HJ91]   L. Hemachandra and S. Jain. On the limitations of locally robust positive reductions. *International Journal of Foundations of Computer Science*, 2(3):237–255, 1991.

[HJ95]   L. Hemaspaandra and Z. Jiang. P-selectivity: Intersections and indices. *Theoretical Computer Science*, 145(1–2):371–380, 1995.

[HJRW97]   L. Hemaspaandra, Z. Jiang, J. Rothe, and O. Watanabe. Polynomial-time multi-selectivity. *Journal of Universal Computer Science*, 3(3):197–229, 1997.

[HJRW98]   L. Hemaspaandra, Z. Jiang, J. Rothe, and O. Watanabe. Boolean operations, joins, and the extended low hierarchy. *Theoretical Computer Science*, 205(1–2):317–327, 1998.

[HJV93]   L. Hemaspaandra, S. Jain, and N. Vereshchagin. Banishing robust Turing completeness. *International Journal of Foundations of Computer Science*, 4(3):245–265, 1993.

[HN93]   A. Hoene and A. Nickelsen. Counting, selecting, and sorting by query-bounded machines. In *Proceedings of the 10th Annual Symposium on Theoretical Aspects of Computer Science*, pages 196–205. Springer-Verlag *Lecture Notes in Computer Science #665*, February 1993.

[HNOS96a]   E. Hemaspaandra, A. Naik, M. Ogihara, and A. Selman. P-selective sets and reducing search to decision vs. self-reducibility. *Journal of Computer and System Sciences*, 53(2):194–209, 1996.

[HNOS96b]   L. Hemaspaandra, A. Naik, M. Ogihara, and A. Selman. Computing solutions uniquely collapses the polynomial hierarchy. *SIAM Journal on Computing*, 25(4):697–708, 1996.

[HNP98]   L. Hemaspaandra, C. Nasipak, and K. Parkins. A note on linear-nondeterminism, linear-sized, Karp–Lipton advice for the P-selective sets. *Journal of Universal Computer Science*, 4(8):670–674, 1998.

[HO02]   L. Hemaspaandra and M. Ogihara. *The Complexity Theory Companion*. Springer-Verlag, 2002.

[HOW02]   L. Hemaspaandra, M. Ogihara, and G. Wechsung. Reducing the number of solutions of NP functions. *Journal of Computer and System Sciences*, 64(2):311–328, 2002.

[HOZZ]   L. Hemaspaandra, M. Ogihara, M. Zaki, and M. Zimand. The complexity of finding and recognizing top-Toda-equivalence-class members. In preparation.

[HPR01]   L. Hemaspaandra, K. Pasanen, and J. Rothe. If P $\neq$ NP then some strongly noninvertible functions are invertible. In *Proceedings of the 13th International Symposium on Fundamentals of Computation Theory*, pages 162–171. Springer-Verlag *Lecture Notes in Computer Science #2138*, August 2001.

[HR99]   L. Hemaspaandra and J. Rothe. Creating strong, total, commutative, associative one-way functions from any one-way function in complexity theory. *Journal of Computer and System Sciences*, 58(3):648–659, 1999.

[HRZ95]   L. Hemaspaandra, A. Ramachandran, and M. Zimand. Worlds to die for. *SIGACT News*, 26(4):5–15, 1995.

[HS65]   J. Hartmanis and R. Stearns. On the computational complexity of algorithms. *Transactions of the American Mathematical Society*, 117(5):285–306, 1965.

[HS01]   S. Homer and A. Selman. *Computability and Complexity Theory*. Springer-Verlag, 2001.

[HT96]   L. Hemaspaandra and L. Torenvliet. Optimal advice. *Theoretical Computer Science*, 154(2):367–377, 1996.

[HW91]   L. Hemachandra and G. Wechsung. Kolmogorov characterizations of complexity classes. *Theoretical Computer Science*, 83:313–322, 1991.

[HZZ96]   L. Hemaspaandra, M. Zaki, and M. Zimand. Polynomial-time semi-rankable sets. In *Journal of Computing and Information*, 2(1), Special Issue: *Proceedings of the 8th International Conference on Computing and Information*, pages 50–67, 1996. CD-ROM ISSN 1201-8511/V2/#1.

[Joc68]   C. Jockusch. Semirecursive sets and positive reducibility. *Transactions of the AMS*, 131(2):420–436, 1968.

[Joc79]   C. Jockusch. Recursion theory: Its generalizations and applications. In *Proceedings of the Logic Colloquium, Leeds*, pages 140–157. Cambridge University Press, 1979.

[JT95]   B. Jenner and J. Torán. Computing functions with parallel queries to NP. *Theoretical Computer Science*, 141(1–2):175–193, 1995. Corrigendum at
            http://www.informatik.uni-ulm.de/abt/ti/Personen/jtpapers.html.

[JY90]   D. Joseph and P. Young. Self-reducibility: Effects of internal structure on computational complexity. In A. Selman, editor, *Complexity Theory Retrospective*, pages 82–107. Springer-Verlag, 1990.

[Kad89]   J. Kadin. $P^{NP[\log n]}$ and sparse Turing-complete sets for NP. *Journal of Computer and System Sciences*, 39(3):282–298, 1989.

[Käm90]   J. Kämper. A result relating disjunctive self-reducibility to P-immunity. *Information Processing Letters*, 33(5):239–242, 1990.

[KL80]   R. Karp and R. Lipton. Some connections between nonuniform and uniform complexity classes. In *Proceedings of the 12th ACM Symposium on Theory of Computing*, pages 302–309. ACM Press, April 1980. An extended version has also appeared as: Turing machines that take advice, *L'Enseignement Mathématique*, 2nd series, 28:191–209, 1982.

[Kle52]   S. Kleene. *Introduction to Metamathematics*. D. van Nostrand Company, Inc., 1952.

[Ko82]   K. Ko. The maximum value problem and NP real numbers. *Journal of Computer and System Sciences*, 24(1):15–35, 1982.

[Ko83]   K. Ko. On self-reducibility and weak P-selectivity. *Journal of Computer and System Sciences*, 26(2):209–221, 1983.

[Köb94]   J. Köbler. Locating P/poly optimally in the extended low hierarchy. *Theoretical Computer Science*, 134(2):263–285, 1994.

[Köb95]   J. Köbler. On the structure of low sets. In *Proceedings of the 10th Structure in Complexity Theory Conference*, pages 246–261. IEEE Computer Society Press, June 1995.

[KS85]   K. Ko and U. Schöning. On circuit-size complexity and the low hierarchy in NP. *SIAM Journal on Computing*, 14(1):41–51, 1985.

[Kum92]   M. Kummer. A proof of Beigel's cardinality conjecture. *Journal of Symbolic Logic*, 57(2):677–681, 1992.

[KW98]   J. Köbler and O. Watanabe. New collapse consequences of NP having small circuits. *SIAM Journal on Computing*, 28(1):311–324, 1998.

[Lan53]   H. Landau. On dominance relations and the structure of animal societies, III: The condition for score structure. *Bulletin of Mathematical Biophysics*, 15(2):143–148, 1953.

[LFKN92]   C. Lund, L. Fortnow, H. Karloff, and N. Nisan. Algebraic methods for interactive proof systems. *Journal of the ACM*, 39(4):859–868, 1992.

[Lon78]   T. Long. *On Some Polynomial Time Reducibilities*. PhD thesis, Purdue University, Lafayette, IN, 1978.

[LS95]   T. Long and M. Sheu. A refinement of the low and high hierarchies. *Mathematical Systems Theory*, 28(4):299–327, 1995.

[MP79]   A. Meyer and M. Paterson. With what frequency are apparently intractable problems difficult? Technical Report MIT/LCS/TM-126, Laboratory for Computer Science, MIT, Cambridge, MA, 1979.

[NRRS98]   A. Naik, J. Rogers, J. Royer, and A. Selman. A hierarchy based on output multiplicity. *Theoretical Computer Science*, 207(1):131–157, 1998.

[NS99]   A. Naik and A. Selman. Adaptive versus nonadaptive queries to NP and P-selective sets. *Computational Complexity*, 8(2):169–187, 1999.

[Ogi94]   M. Ogihara. Polynomial-time membership comparable sets. In *Proceedings of the 9th Structure in Complexity Theory Conference*, pages 2–11. IEEE Computer Society Press, June/July 1994.

[Ogi95]   M. Ogihara. Polynomial-time membership comparable sets. *SIAM Journal on Computing*, 24(5):1068–1081, 1995.

[Ogi96]   M. Ogihara. Functions computable with limited access to NP. *Information Processing Letters*, 58(1):35–38, 1996.

[Pap94]   C. Papadimitriou. *Computational Complexity*. Addison-Wesley, 1994.

[Sch76]   C. Schnorr. Optimal algorithms for self-reducible problems. In *Proceedings of the 3rd International Colloquium on Automata, Languages, and Programming*, pages 322–337. Edinburgh University Press, July 1976.

[Sch83]   U. Schöning. A low and a high hierarchy within NP. *Journal of Computer and System Sciences*, 27(1):14–28, 1983.

[Sch86]   U. Schöning. Complete sets and closeness to complexity classes. *Mathematical Systems Theory*, 19(1):29–42, 1986.

[Sel74]   A. Selman. On the structure of NP. *Notices of the AMS*, 21(5):A–498, 1974. Erratum in the same journal, 21(6):310.

[Sel78]   A. Selman. Polynomial time enumeration reducibility. *SIAM Journal on Computing*, 7(4):440–457, 1978.

[Sel79]   A. Selman. P-selective sets, tally languages, and the behavior of polynomial time reducibilities on NP. *Mathematical Systems Theory*, 13(1):55–65, 1979.

[Sel81]   A. Selman. Some observations on NP real numbers and P-selective sets. *Journal of Computer and System Sciences*, 23(3):326–332, 1981.

[Sel82a]   A. Selman. Analogues of semirecursive sets and effective reducibilities to the study of NP complexity. *Information and Control*, 52(1):36–51, 1982.

[Sel82b]   A. Selman. Reductions on NP and P-selective sets. *Theoretical Computer Science*, 19(3):287–304, 1982.

[Sel88]   A. Selman. Promise problems complete for complexity classes. *Information and Computation*, 78:87–98, 1988.

[Sel94]   A. Selman. A taxonomy of complexity classes of functions. *Journal of Computer and System Sciences*, 48(2):357–381, 1994.

[Sha92]   A. Shamir. IP = PSPACE. *Journal of the ACM*, 39(4):869–877, 1992.

[Sip97]   M. Sipser. *Introduction to the Theory of Computation*. PWS Publishing Company, 1997.

[Siv98]   D. Sivakumar, December 1998. Personal communication.

[Siv99]   D. Sivakumar. On membership comparable sets. *Journal of Computer and System Sciences*, 59(2):270–280, 1999.

[SL94]   M. Sheu and T. Long. The extended low hierarchy is an infinite hierarchy. *SIAM Journal on Computing*, 23(3):488–509, 1994.

[Tan00]   T. Tantau.   On the power of extra queries to selective languages.
            Technical Report TR00-077, Electronic Colloquium on Computational
            Complexity, http://www.eccc.uni-trier.de/eccc/, September 2000.
[Tan01]   Till Tantau, September 2001. Personal communication.
[Tan02]   T. Tantau. A note on the power of extra queries to membership com-
            parable sets. Technical Report TR02-004, Electronic Colloquium on
            Computational Complexity, http://www.eccc.uni-trier.de/eccc/, Jan-
            uary 2002.
[Tod91]   S. Toda. On polynomial-time truth-table reducibilities of intractable
            sets to P-selective sets. *Mathematical Systems Theory*, 24(2):69–82,
            1991.
[Ver94]   N. Vereshchagin, January 1994. Personal communication.
[Wan95]   J. Wang. Some results on selectivity and self-reducibility. *Information
            Processing Letters*, 55(2):81–87, 1995.
[Wes96]   D. West. *Introduction to Graph Theory*. Prentice-Hall, 1996.

Index

Monographs in Theoretical Computer Science · An EATCS Series

Texts in Theoretical Computer Science · An EATCS Series